AGRICULTURAL AND FOOD CONTROVERSIES

AGRICULTURAL AND FOOD CONTROVERSIES

WHAT EVERYONE NEEDS TO KNOW®

**F. BAILEY NORWOOD,
PASCAL A. OLTENACU,
MICHELLE S. CALVO-LORENZO,
AND SARAH LANCASTER**

OXFORD
UNIVERSITY PRESS

OXFORD
UNIVERSITY PRESS

Oxford University Press is a department of the University of Oxford.
It furthers the University's objective of excellence in research, scholarship,
and education by publishing worldwide.

Oxford   New York
Auckland   Cape Town   Dar es Salaam   Hong Kong   Karachi
Kuala Lumpur   Madrid   Melbourne   Mexico City   Nairobi
New Delhi   Shanghai   Taipei   Toronto

With offices in
Argentina   Austria   Brazil   Chile   Czech Republic   France   Greece
Guatemala   Hungary   Italy   Japan   Poland   Portugal   Singapore
South Korea   Switzerland   Thailand   Turkey   Ukraine   Vietnam

Oxford is a registered trademark of Oxford University Press
in the UK and certain other countries.

Published in the United States of America by
Oxford University Press
198 Madison Avenue, New York, NY 10016

Library of Congress Cataloging-in-Publication Data
Norwood, F. Bailey.
Agricultural and food controversies : what everyone needs to
know / F. Bailey Norwood, Pascal A. Oltenacu, Michelle S.
Calvo-Lorenzo, and Sarah Lancaster.
pages cm
Includes bibliographical references and index.
ISBN 978-0-19-936843-3 (hardcover : alk. paper) —
ISBN 978-0-19-936842-6 (pbk. : alk. paper)   1. Agriculture—Moral and
ethical aspects.   2. Food—Moral and ethical aspects.   I. Title.
BJ52.5.N67 2014
174'.963—dc23
2014024765

1 3 5 7 9 8 6 4 2
Printed in the United States of America
on acid-free paper

CONTENTS

ACKNOWLEDGMENTS

Much of the information in this book was acquired through the normal course of our jobs, which are funded by the US Department of Agriculture, the state of Oklahoma, student tuition, and various federal and state grants. All of these have our gratitude.

We also had the opportunity to learn new things while writing the book, and were enlightened by numerous conversations with fellow agricultural scientists and government employees at the US Department of Agriculture, the Environmental Protection Agency, and the Food and Drug Administration. To better understand the nature of food controversies we reached out to individuals at activist and nonprofit organizations, writers, documentary producers, farmers, and the agribusiness and food industry. Many such individuals were willing to engage in thoughtful conversations and review parts of the manuscript, and we appreciate their help.

Writing this book required us to pause from our normal routine of writing esoteric journal articles for fellow scientists and engage the general public, something we could not do without the enthusiastic support of our department heads and deans at our universities.

Note

A complete list of all references and sources is available at fbaileynorwood.com.

Readers are welcome to contact any of the authors regarding questions or comments. Norwood and Oltenacu contributed to all chapters. Lancaster contributed to chapters 2, 3, and 4; and Calvo contributed to the animal welfare portion of chapter 8.

AGRICULTURAL AND FOOD CONTROVERSIES

1

WHAT I EAT IS YOUR BUSINESS

Jonathan Haidt is a social psychologist who studies the different values held by liberals and conservatives. He writes in his 2012 book, *The Righteous Mind*, "Liberals sometimes say that religious conservatives are sexual prudes....But conservatives can just as well make fun of liberal struggles to choose a balanced breakfast—balanced among moral concerns about free-range eggs, fair-trade coffee, naturalness, and a variety of toxins, some of which (such as genetically modified corn and soybeans) pose a greater threat spiritually than biologically" (13).

A friend of ours who is also a food activist remarked in response, "And conservatives don't care what they put in their bodies as long as it is quick, convenient, and cheap!" Both comments were made in humor, as their authors are not political pundits, and neither of them wish to be affiliated with any one political party. Yet there is always some truth to humor, and food has indeed become a politically divisive topic.

People have always talked about food, but in the past it was largely in regard to personal health, religion, taste, and affordability. Now food is also a public issue in that what you eat impacts you and all of society, making agriculture an ethical issue. If you doubt this, go to Amazon Instant Video and search the term "food documentary." There you will find more than eighteen documentaries questioning how our food is raised, with suggestions on how to make it more ethical.

How a crop is raised influences the amount of soil lost to erosion, whether lakes are polluted, emissions of greenhouse gases, and the ability of future generations to feed themselves. Livestock are sentient creatures, and consumers want to know their food is humanely raised. Because how you eat affects other humans and animals, your fellow citizens are keen to make sure you are eating what *they* consider to be ethical foods. Farmers want to produce ethical food as well. The problem is that there is considerable disagreement about what "ethical" food is. What you eat used to be your business; now it is everyone's business—little surprise that agriculture has become such a controversial subject!

These food fights can get nasty, like when Jon Stossel called New York state representative Felix Ortiz a "cancer" for wanting to tax junk food (by the way, Ortiz responded that he is a "good cancer"), or when Robert F. Kennedy called hog farmers a bigger threat than Osama bin Laden. As the outlandish insults fly, so does the money, as each side seeks to lobby harder than the other. We wrote this book because we felt there was too much name-calling, and too many books and documentaries representing only one side of the debate. Our research in agricultural economics has given us the unique opportunity to interact with industry and interest groups, and we have learned that both sides consist of smart, kind people wishing to produce healthy, affordable food in an ethical manner.

Controversial subjects can be explored while paying respect to the character and intellect of both sides, and we seek to do so in this book. As we tour the gallery of agricultural controversies, we will try to illustrate why equally smart and kind people can form vastly different opinions about food, and then provide our perspectives on what the economic and scientific literature says about the issues. The idea is not to convince readers to adopt our perspectives, or to declare one side of a food debate as champion, but to help readers reach informed opinions, whatever those opinions may be.

For a preview of what divides people, consider a recent Gallup poll that asked Americans whether they have a positive or negative image of free enterprise. The vast majority (88–94%) of both Democrats and Republicans viewed it positively (see figure 1.1). Americans of both political camps apparently respect the pursuit of an honest buck. Of course, Republicans and Democrats do not agree on everything, and the Gallup poll discovered that Democrats and Republicans view large institutions, like big business and big government, differently. Whereas 75 percent of Republicans view big business positively, only 44 percent of Democrats feel likewise. When asked about the federal government, most (75%) of Democrats view it favorably, compared to only 27 percent of Republicans. The data are clear: Republicans dislike big government and Democrats dislike big business (or, at least, that is what they say), and as we will see, some controversies have just as much to do with attitudes towards large corporations as they do the science of farming. What isn't up for debate is

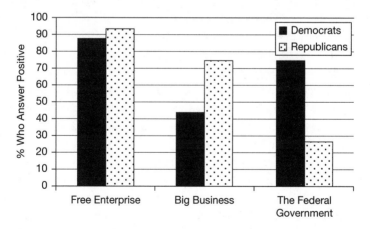

Figure 1.1 Political Ideologies in the United States

Note that "Democrats" refers both to people who call themselves Democrats and those who lean towards the Democrats. A similar statement can be made about the "Republicans."

Source: Frank Newport, "Democrats, Republicans Diverge on Capitalism, Federal Gov't," *Gallup Politics*, November 29, 2012.

the free-enterprise system of food production—that should be made clear. People aren't debating capitalism versus socialism, but what type of regulated capitalism is best. As the reader probably suspects, liberals prefer more government regulation when it comes to food health and food safety. Consumers of organic food and supporters of animal welfare legislation are also more likely to lean to the political left. It is not a stretch to say that food activists—by which we mean the most vocal individuals seeking changes in agriculture—are mostly liberals. (The term "food activist" is not meant as a euphemism, nor is it intended to suggest an extremist, but is used to reflect the passion with which some advocate for changes in agriculture.) Now combine two facts: (1) liberals have a negative view of big business, and (2) liberals comprise the majority of food activists, and you have a story that can explain the rise of many agricultural controversies. It is an overly simple story, as most people have more nuanced views than this story would suggest. For instance, views about genetically modified foods cannot be easily explained by political affiliation, and some evidence suggests that conservatives express greater disapproval of genetically modified foods than liberals.

Then again, basically every county in California that voted to reelect President Obama also voted in favor of labeling of genetically modified food, with the reverse being true in other counties, so when it comes to genetically modified foods, it is regulation that is controversial. You see, agricultural controversies are not just scientific controversies but have a political component as well. Most scientists would prefer to keep agricultural science and politics separate, but these days, politics and food go together like salt and pepper. This is evident in where we buy food. If you drive through a region populated with an unusually large number of Cracker Barrel restaurants, studies have shown that the region is probably dominated by Republicans. Likewise, Whole Foods grocery stores tend to prosper in Democratically controlled districts. However uncomfortable politics may be, agricultural debates cannot be discussed honestly without including politics. To ignore political issues about food is to dismiss as irrelevant those who make political arguments, and this book endeavors to take all arguments—and all people—seriously.

Don't worry, this is not a book about conservatives versus liberals. What is most important in explaining the political side of agricultural controversies is not one's political party, but one's attitude towards large corporations. It is striking how often the world "corporation" appears in books and documentaries by food activists. For this reason, the words "conservative" and "liberal" will not appear in subsequent chapters, but the term "corporation" will be used throughout.

It is interesting that liberals dislike big business and conservatives dislike big government, because agriculture in modern democracies consists of both. Agriculture used to consist mostly of small farmers, small craftsmen supplying their inputs, and small businesses distributing food to consumers. From the Middle Ages to the early nineteenth century about 90 percent of the population labored on farms. Today that percentage is less than 2 percent. Agricultural production has not fallen, though. Amazingly, it has risen because the average farm has increased in size, and more importantly, soared in productivity. This rise in productivity is partially attributable to the dramatic increase in efficiency and innovative technology fostered in part by what we today call "agribusiness." Chemical fertilizers, synthetic pesticides, synthetic growth hormones, and better crop and livestock genetics have increased the amount of food each farmer can produce.

The vast majority of our food has passed through at least one large corporation between the farm and the fork, and for those who distrust big business, this fact can create an atmosphere of suspicion. Why are large farms and corporations so dominant in food? One reason is economies of scale, whereby a firm can produce each unit at a lower cost the more of those units it produces. Studies have shown that large Illinois farms of 900 soybean acres experience production costs 82 percent (per bushel) lower than 300-acre farms. Large corn farms produce at a 38 percent lower cost (per bushel) than small corn farms. Likewise, dairies with more than 2,000 head of cattle produce at a lower cost (per gallon of milk) than dairies with 30 cows or less. Large hog-slaughtering facilities can process hogs at a cost 11 percent lower (per lb.) than small facilities. A large brewery has half the costs (per ounce) of a small one.

Also, large corporations can afford the research and development costs necessary to invent and market scientific technologies like pesticides, chemical fertilizers, and genetically modified crops. It is largely because of economies of scale and new technologies that world food prices have steadily fallen in the last hundred years, even while the world population has risen and the number of farmers has fallen.

Food activists do not contest the numbers in figure 1.2, but they insist that the quality of food has also fallen, and that

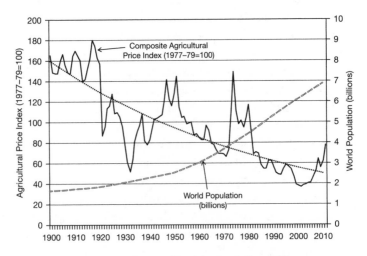

Figure 1.2 Agricultural Product Prices and Population Growth Since 1900

Source: Keith Fuglie and Sun Ling Wang, "New Evidence Points to Robust but Uneven Productivity Growth in Global Agriculture," *Amber Waves*, September 20, 2012. Economic Research Service, US Department of Agriculture. Data for chart provided by Keith Fuglie on August 15, 2013.

industrial agriculture externalizes some of its costs onto society, making the real cost of food higher than the price in the grocery store. For example, it may get away with polluting the water, leaving the cost of the cleanup to others.

Activists sometimes argue that corporations grow big not just to benefit from economies of scale, but to gain market power and political influence. They look at the endless varieties of food in the grocery store but see only a few corporations producing it all, making them feel like they are at the mercy of big business. Figure 1.3 depicts how a large number of food brands can be produced by only a few corporations, making it difficult to assess whether the food market is competitive or not. To counter the power of big business, modern democracies also evolved big government, manifested in the many regulations regarding farming and food processing. Food safety laws can be so onerous that they prohibit an individual from giving free food to the homeless. Regulations have their benefits also, and these regulations help to ensure that food is not adulterated, pesticides approved for sale are safe, lakes are protected from fertilizer runoff, meat does not contain traces of antibiotics, and livestock are slaughtered humanely.

The rise of big business and big government can be a good thing, simultaneously allowing economies of scale to lower the price of food and regulations to protect us from irresponsible corporate behavior. Food activists appear to take the opposite view though, believing that what has really resulted is big business corrupting big government, allowing corporations to write their own rules. When food activist and best-selling author Michael Pollan appeared on *The Colbert Report* in 2013, he suggested food cooked by a corporation is unhealthy. The Cornucopia Institute has published a diagram titled *Is the USDA a Wholly-Owned Subsidiary of Monsanto?* listing fifteen individuals who have held important positions in both the USDA and Monsanto. The organization Food Democracy Now! describes itself as being motivated by the fact that the US government cares more about the interests of corporate agribusiness than

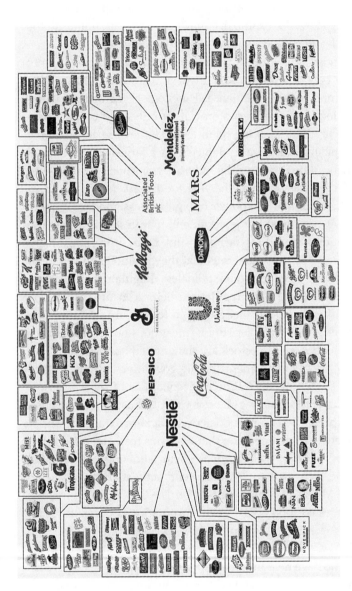

Figure 1.3 Is the Food Market Competitive?

Source: Joki Desnommée-Gauthier, 2012. Figure created for Oxfam International. Accessed September 24, 2013, at http://firstperson.oxfamamerica.org/files/2013/02/graphic-72dpi-800px-english.png.

about farm families and consumers. Of course, this is no new claim, and is not specific to just agriculture.

Some activists believe the twin towers of big business and big government are unavoidable, and thus support an even larger government, hoping that it will exert more democratic control over food production. The Food Democracy Now! organization, mentioned earlier, believes that corporations dictate government policies, yet at the same time is asking government to make labeling of genetically modified food mandatory. Another example is the Humane Society of the United States, whose strategy assumes that large, confined animal-feeding operations are here to stay (yes, the Humane Society would like a worldwide conversion to veganism, but does not believe it likely) and thus pursues regulations to reduce the suffering it believes farm animals experience.

Unsatisfied with big business and big government, some are asking us to think smaller, suggesting we should obtain food from farms that are the antithesis of big business. Organic farming emerged as a desire to do without the fertilizers and pesticides produced in large factories, or the seeds produced in a laboratory. Organic food is partially a protest against the industrial style of production that is so prevalent in the modern economy, but thought by some to be incompatible with ethical food. Then Walmart started selling organic food, and for some, organic lost its allure. It "sold out" to big business, one might say.

Walmart is successful for many reasons, one being its large distribution system connecting farmers and consumers hundreds of miles away. That system is not equipped to sell local foods, it would seem, so when locavores began writing books and producing films, they considered local foods immune to competition from corporations. Never underestimate Walmart though, as it eventually figured out how to compete in this market as well. Modern food movements are akin to a game of catch, whereby food activists seek to distinguish themselves from big business, only to have corporations co-opt their cause.

Food controversies are as much about who sells the food as how the food is grown. This is evident in the genetically modified food controversy, where the groups leading the opposition seem to dislike the corporation Monsanto more than the technology itself. Type "Monsanto" into a Google search and it will sometimes suggest you add the word "evil," because that is what many other users have done. Readers of NaturalNews .com in a 2011 online survey even voted Monsanto the "Most Evil Corporation of the Year."

Food activists are not just leading a "small is beautiful" movement or a "more regulation" movement; they are pursuing a change in food culture. They are the Occupy Wall Street of food in that they wish to inject more democracy into food, not through a vertical top-down system of political power, but a horizontal, informal network of concerned citizens who ask at each point in the food channel, "How does this affect society at large, animals, and the environment?" They ask a lot of questions. They write books. They form organizations, and websites. They produce food documentaries. And when they feel it necessary, they will lobby for laws to oppose corporations—and when they do, they name their organization things like "Food Democracy Now!" They lobby because corporations lobby, and the arms race for political influence wages on.

The offspring of this new food movement are not just new products to buy at the grocery store, but new questions to ask about food. We are not just asked to buy organic food and support more regulations, but to think of the soil differently, to be more mindful of our carbon footprint, and to consider the emotions of farm animals. Consumers, food activists, farmers, and the food industry are asking profound questions about food, and the questions are worthy of our attention.

These controversies concern chemical fertilizers, pesticides, global warming, genetically modified organisms, farm subsidies, market power, local foods, and how we raise livestock. Each controversy can be approached in a variety of ways. We as authors, choose to take on the issues as they are fought in developed nations, and mostly the United States, not because they are necessarily more important, but because they are what we are most familiar with. Much of the developing world just wants to feed its people and raise enough cash crops to help their economies grow out of subsistence and into the affluent world. The needy people of the developing world are likely puzzled as to why some Americans want to pay higher food prices. For some in the United States, western Europe, and a few other locations like Australia, food is not just the fuel of life—it is part of their identity. The foods they buy at the market and the restaurants they patronize signal their beliefs and values. We all wish to contribute to society in some fashion, and some choose food as their altruistic outlet. The affluent world has the luxury to pay more attention to the environment and animal welfare, and as the Third World follows, it may do the same. This means that the agricultural controversies we discuss are relevant to both the developed world today and the developing world tomorrow.

2

THE PESTICIDE CONTROVERSY

What Is the Pesticide Controversy?

Margot Woelk was ninety-five years old before she revealed her role in Nazi Germany as Hitler's food taster. Fearful the British would poison him, Hitler made sure to eat food only after it was eaten by Margot and fourteen other girls serving as his official tasters. Hitler may have been evil but he was not stupid. He knew that poisons affect people differently, and knew that any food that harmed one girl might harm him (then pity what would happen to the cook!).

Every year we spray something akin to poison on our food, and use something akin to Hitler's system of making sure we are not harmed. The motives are polar opposites—Hitler cared only for the preservation of his person, while we seek the safety of all humans. Whether they are synthetic pesticides or "natural" pesticides used in organic food, they are applied with the intention of killing three types of pests: insects, weeds, and pathogens (e.g., fungi and viruses). At some level they could poison us also. Many contain carcinogens, cause neurological disorders, and the like. Yet our food seems safe to most people, and since 1992 cancer incidence rates have fallen or remained the same, cancer death rates have fallen, and life expectancy in the United States has been steadily increasing.

Can we be absolutely sure pesticides are used safely? Not entirely, but like Hitler (and according to movies, every Roman emperor, every pope, and every medieval king) we employ testers—not in the form of humans, but animals. All pesticides must be approved by the Environmental Protection Agency (EPA), where the pesticide under consideration is given to laboratory animals at different dosages. The animals' health is monitored over time and used to gauge the threats to human health a pesticide may pose. The EPA then determines whether the pesticide should be allowed and, if it is, the specific instructions on how it should be applied.

Is it cruel to test pesticides on animals? It certainly isn't something we enjoy doing, but not to test on animals will cause us to harm humans—a notion on which 90 percent of toxicologists agree. Pesticides decrease the cost of food, and make fruits and vegetables more affordable. Raise the price of these healthy foods and cancer rates and other health problems in humans will rise. Help the lab animals, and you harm some humans. Modern, democratic societies must make a trade-off between harm to laboratory animals and harm to humans. In a sense, we must "pick our poison," while trying to make the overall harm to animals and humans as low as possible.

Hitler was willing to sacrifice fifteen girls to save himself. The modern world is willing to sacrifice a small number of laboratory animals to protect millions of humans. Moreover, the EPA continues to find ways to reduce testing on animals without sacrificing food safety, like recent developments in molecular and computational sciences, which can sometimes be substituted for animal experimentation.

In June 2013 *The Wall Street Journal* hosted a debate titled "Would Americans Be Better Off Eating a Mostly Organic Diet?" and it tended to center on pesticides. It featured one person who answered yes and one who answered no, to the question in the title, and the justification for their answers describes the pesticide controversy nicely. One person argued in favor of organic foods under the belief that regulatory agencies do an inadequate job of protecting public health, and the other argued that conventional food is not only safe, but also that the use of pesticides makes fruits and vegetables more affordable.

> LU (ALEX) CHENSHENG: Many say the pesticides found in our food are nothing to fear because the levels fall well below federal safety guidelines and thus aren't danger-ous....But federal guidelines don't take into account what effect repeated exposure to low levels of chemicals might have on humans over time. And many pesticides were eventually banned or restricted by the federal gov-ernment after years of use when they were discovered to be harmful to the environment or human health.
>
> JANET H. SILVERSTEIN: Given the lack of data showing that organic food leads to better health, it would be counter-productive to encourage people to adopt an organic diet if they end up buying less produce as a result....As for pesticide exposure, the U.S. in 1996 estab-lished maximum permissible levels for pesticide resi-dues in food to ensure food safety. Many studies have shown that pesticide levels in conventional produce fall well below those guidelines.
>
> —"Would Americans Be Better Off Eating a Mostly Organic Diet?" *Wall Street Journal*, June 17, 2013. R3.

The pesticide controversy boils down to whether the regula-tory agencies are making wise decisions about how pesticides are used or whether we must take measures to protect our-selves. In the United States, that agency is the Environmental Protection Agency (EPA), and it is charged with permitting pesticides only when they do not present an unreasonable risk to humans or the environment, while also taking into account the economic costs and benefits. The controversy is whether the EPA fulfills this charge.

What Are the Benefits and Harms of Pesticide Use?

Before delving into the regulation of pesticides we must develop a better appreciation of the benefits and potential harms of pesticides. The benefits are that they protect crops from damage by insects, weeds, and pathogens, allowing farmers to produce more food using the same amount of inputs. For consumers, this means greater availability of foods and lower prices.

Peanuts are one of the healthiest foods and are relatively inexpensive. If no pesticides were allowed, peanut yields would fall by 78 percent; about one-third of this reduction would be due to the absence of herbicides and two-thirds to the absence of insecticides and fungicides combined. As fewer peanuts were sold on the market, prices would rise, probably by about 150 percent. Rice is a staple food for much of the world, and without pesticides yields would fall by 57 percent. If they were grown without pesticides, the yield for some of our healthiest foods, like apples, lettuce, tomatoes, and oranges, would fall by more than 50 percent (all are US numbers). These are the same fruits and vegetables experts keep telling us to eat in greater portions. Pesticides allow us to produce the same amount of food using less land, and make it easier for farmers to employ no-tillage farming techniques where no plowing is performed, thereby reducing soil erosion and fertilizer runoff. Many of the genetically modified crops today are valued because of their resistance to pesticides, but we defer this issue to another chapter.

A Chinese cook recently demonstrated the potential harms of pesticides when he mistook a pesticide for a spice. One person died and twenty others were sickened. Pesticides per se are not poisons though. The First Law of Toxicology, established in the sixteenth century, is that it is the dose, not the chemical, that makes a poison. We are constantly exposed to natural pesticides in our daily life. After all, plants make their own pesticides to ward away pests, and we eat many of these plants.

If people are exposed to them at unsafe dosages, pesticides can cause cancer and a variety of neurological disorders, like

Parkinson's disease. To what extent has pesticide use over the last few decades harmed human health? The more we learn, the more difficult it is to say. In the early 1980s research concluded that pesticides played a very minor role in human health problems, leading some to conclude that virtually nobody dies of cancer caused by pesticides. Since then we have learned how difficult it is to determine the impact of pesticides on health, given the variety of carcinogens we encounter (including charred meat, acrylamide in French fries and coffee, and household cleaning supplies) and the long delay between exposure and health impacts. Scientists are fairly certain that about one-third of cancer is caused by smoking and another one-third is caused by poor diet, overweight, and too little exercise, but the sources of the remaining third are difficult to assign.

Of this other third of cancers, pesticide use certainly seems to play some role. Non-Hodgkin's lymphoma, prostate cancer, melanoma, and a variety of other cancers are correlated with pesticide use. People applying pesticides, living on farms, or employed in pesticide manufacturing seem to have higher cancer rates than people who rarely encounter pesticides.

The issue becomes even more complex when one considers the many indirect ways pesticides affect humans. Honeybee colonies have reduced dramatically in recent years in something called colony collapse disorder, and though the cause isn't certain, pesticides could be partly to blame. Since we rely on bees to pollinate much of our fruits and vegetables, this indirect effect could negate any direct benefits of certain pesticides.

There is little controversy over whether pesticides may pose a potential harm. What is questionable is whether actual harms are observable, and if they are, whether the benefits of pesticides outweigh those health harms. For instance, a pesticide may directly increase cancer rates slightly, but indirectly cause a larger reduction in cancer rates by reducing substantially the price of fruits and vegetables. When the Mayo Clinic listed seven tips to reducing risk of cancer, the first tip was to abstain from tobacco and the second was to eat a healthy diet, which was described as lots of fruits and vegetables, a limited amount of fat, and avoiding too much alcohol. Avoiding foods produced using pesticides was not even on the list.

Now that we recognize this trade-off between pesticide harms and benefits, we turn to the regulation of pesticides in Western democracies, focusing mostly on the US regulatory system. While the legal framework for regulating pesticides differs in western Europe, the methods, challenges, and goals are very similar. Much of what is said about the EPA can be extrapolated to the European Union and the United Kingdom.

How Are Pesticides Regulated?

It is not unusual to hear about salespeople in the early days of synthetic pesticides (1940s) who would drink the chemical to prove its safety. One always suspects the salesmen were playing a ruse, but it is a testimony to how safe people once considered pesticides. The pesticide DDT was called a "savior of mankind" during World War II, as it was the first war where more people died from wounds than from disease. Farmers began using DDT on a large scale, and governments would spray generous amounts on bodies of water to kill mosquitoes.

Rachel Carson was not so impressed though, as she began to document the cumulative effect of DDT in animals. In 1962, she published her scathing indictment of DDT in her book *Silent Spring*. This book launched an environmental movement that continues today. Her book is widely credited with convincing President Richard Nixon to establish by executive order the Environmental Protection Agency eight years later. The EPA acknowledges in its official history that it was *Silent Spring* that prompted the federal government to address the threat of pesticides, along with other environmental problems.

Pesticides have been used since ancient times. In *The Odyssey*, Homer has Ulysses bellow to his nurse, "Bring blast-averting sulfur, nurse, bring fire! / That I may fumigate my walls." It is likely that the Greeks had used sulfur since prehistoric times, and that experience taught them how to use it safely. Today synthetic pesticides are typically created in a factory. New formulations are continually introduced, ones humans do not have generations of experience using, so controlled experiments are needed to determine what health threat they may pose.

The United States requires all pesticides to be registered with the EPA, and older pesticides are continually reviewed to make sure they meet the newer safety requirements. When a pesticide is registered, it can then be used, but only in settings and at levels approved by the EPA. If the EPA makes wise decisions about registering pesticides and determining approved dosages, then little to no harm should come from pesticide use.

To determine whether a pesticide is safe the EPA first requires the pesticide company to provide data regarding the largest amount of pesticide residues one would expect to see on the crops in the field (when pesticides are applied at their highest dosage) and in processed food made from those crops. The agency then seeks to determine if those residues are harmful. This is where the tasters—laboratory animals—are used. By exposing animals to different levels of the pesticides, researchers can determine the threshold beyond which they cause harm to the animals. This threshold can be stated in terms of residues divided by the animal's weight, so that it can be used to determine the appropriate threshold for humans.

In toxicology this threshold may be specified as a median lethal dose, or LD_{50}, which refers to the dose required to kill half of the animals exposed in experiments. It is a standardized dosage that allows us to compare the relative dangers posed by different chemicals, and in doing so it sometimes shows how safe many pesticides are. The herbicide glyphosate used on almost all soybean acres has an LD_{50} of 4,320 milligrams. This seems safer than table salt (LD_{50} = 3,300 mg) and much safer than caffeine (LD_{50} = 192 mg). If you do not fear the caffeine in your coffee, then there seems little to fear from the herbicides applied to soybeans.

Measures like the LD_{50} are mostly used to determine the potential hazard to farmworkers applying the pesticides. To determine potential risks in food consumption, the EPA doesn't use LD_{50} as a measure but something referred to as the "No Observable Adverse Effects Level," or NOAEL. This is the highest dose of a pesticide that results in no negative response in the animal, and that negative response could be almost anything, including weight loss or changes in the body's production of an enzyme. These studies are so comprehensive they sometimes observe animals over multiple generations.

Human biology is not the same as that of lab animals, so to be extra safe, that NOAEL threshold (again, in units like residues per pound) is then divided by a "safety" factor—a large number from 100 to 1,000—so that the EPA is comfortable deeming the pesticide as safe. This threshold takes into account all the avenues by which residues may reach the consumer, so it considers the total diet of consumers, including food imports and even drinking water.

So pesticides are only expected to harm humans when they are exposed to a dosage a hundred or a thousand times larger than the dosage observed to harm animals. To understand the importance of this safety factor, try this experiment. Consume large portions of chocolate in one day—more than you ever imagined eating in your life. Chances are that you will be okay. Then feed a dog the same amount of chocolate per pound of weight—actually, don't do that, as the dog would probably die. This is why the EPA uses such a large safety factor. If you fed a dog one one-hundredth as much chocolate as you ate, it would probably be okay.

The bodies of infants and children react differently to pesticides, so other factors must be considered to protect kids. For instance, the Food Quality Protection Act states that if reliable data on threshold effects for a child are not available, the safety factor should be increased by a factor of ten, perhaps increasing from 1,000 to 10,000.

Why must we experiment on animals? Because controlled experiments are absolutely necessary for determining when a pesticide *causes* health harms. In the real world, greater exposure to pesticides may be correlated with poor health, but the correlation may not be causation. Someone who eats nonorganic food may also tend to eat fewer vegetables, smoke, and rarely exercise. If those people are more likely to develop cancer, was it the pesticides that caused it? Or was it too few vegetables, or insufficient exercise? One cannot tell, and so controlled experiments are necessary for determining what happens to an animal when pesticide use increases but everything else stays the same. They are so necessary that around 90 percent of toxicologists disagree with the statement "Animal testing is not needed."

This threshold mostly relates to the prevention of noncancer health problems. If a pesticide is shown to cause cancer in laboratory animals when given in high doses, the EPA will assume there is no safe dosage, and the pesticide is denied registration. The EPA certainly is not lax when it comes to allowing pesticides to be applied and generally will not approve a pesticide if it increases people's risk of having cancer by even one in one million.

Regulators don't just measure the potential harms to humans but to the environmental as well. The EPA considers a broad array of environmental impacts and even assesses the potential harm to threatened and endangered species. When the neonic class of pesticides was approved for use, it could not have been anticipated that they might cause a collapse in bee colonies. Later, when research determined they might be partly responsible, the European Union placed a two-year ban on their use, and the EPA is studying the situation to see if new restrictions are desirable.

Pesticide regulation does not just take into account the safety of a pesticide but its benefits also. A chemical can directly harm humans through exposure but can benefit human health by keeping the price of healthy foods low—especially prices of fruits and vegetables. Thus a pesticide with a low NOAEL may pose less harm than one with a higher NOAEL if it does a better job of providing affordable fruits and vegetables. The EPA would be remiss if it did not consider the benefits of a pesticide on farm productivity when articulating how it should be used.

Finally, regulation does not stop with the animal trials. Humans may respond differently to pesticides than animals do, and there is no guarantee that the safety factors used offer enough protection. Also, experiments cannot reveal the cumulative danger of exposure to all the pesticides that are used. It's like drinking one sip out of many, many bottles of wine. Each sip has only a negligible effect on your ability to drive, but when the sips are added together, you do not belong behind the wheel. Researchers are constantly collecting data on the health of individuals and their exposure to toxic chemicals like pesticides, to detect any alarming correlations. This field of research is called epidemiology, and it serves as a second opinion on the effectiveness of pesticide regulations. Epidemiological studies are used to revise established regulations and to help the government develop better guidelines on the regulations of new pesticides in the future.

How Effective Are Pesticide Regulations?

It should be apparent by now that the EPA and its European counterparts set high safety standards regarding pesticides based on controlled animal experiments and epidemiological studies. The question is whether those standards are met. If pesticides only impact humans as they do animals in experiments, and if pesticide regulations are properly enforced, then the use of pesticides in agriculture is very safe. Safe use of pesticides is possible today partly because new technologies can detect residues at around one part per quadrillion (like detecting a grain of salt in an Olympic-sized swimming pool!). To illustrate, you would have to eat more than 7,000 tomatoes per day throughout your life to reach the maximum residue level of pesticides inherent in conventional tomatoes. Since you eat far, far less than this, there is no reason to fear conventional tomatoes.

Government agencies sample and check foods to ensure tolerance levels are being observed, and for the most part they are. Of the grain, dairy, seafood, and fruits sampled in 2008 none displayed residue levels above EPA's tolerance level. Only 1.7 percent of vegetables exceeded the tolerance level. The numbers were slightly higher for imported food, though still less than 5 percent (save for food group "other" at 8.3 percent). Other studies support this finding that pesticide residues only rarely exceed the EPA maximum. Remember, even the rare food that does exceed the limit contains pesticides at a far lower level than that which causes health problems in laboratory animals.

Epidemiological studies, however, do find that pesticides impact human health. For three years one of the authors has printed and filed almost every article about pesticides from ScienceDaily.com. What percentage of these articles find that pesticides harm human health? Almost 100 percent! One says that prenatal exposure to DDT causes high blood pressure later in life. Another suggests a link between the pesticide benomyl and Parkinson's disease. And another links the pesticide additive PBO with noninfectious coughing of young children. There are many others (to see for yourself, go to ScienceDaily. com and just search for the word "pesticide").

The problem with epidemiological studies is that it is very easy to establish correlations between health impacts, food, and the environment, but establishing causation is impossible. If consumers who eat organic food and consume less pesticide residues also tend to eat healthier foods and exercise more, and one finds these individuals have lower cancer rates, how can you tell whether the cancer reduction was caused by less pesticides, better food, or more exercise?

Suppose for argument's sake that correlation did mean causation. Could it really be that every single epidemiological study finds a link between pesticide use and health problems? No, but only those studies that do find a link are deemed interesting enough to publish. Would you read an article titled "Use of Popular Pesticide Not Linked to Health Problems"? What about an article titled "Popular Pesticide Shown to Cause Infant Death, Early Onset of Parkinson's Disease, and Brain Cancer"? Both academic and popular publishers know the answer, and are consequently more likely to publish the second article and reject the first. Only the researchers who know about both published and unpublished studies know a pesticide's true impact.

In the end, as with many agricultural controversies, opinions about the use of pesticides often boil down to whether regulators are making wise judgments. Wise judgments require experience, knowledge, and also the proper incentives. If one believes that politicians, regulatory agencies, and pesticide corporations are rife with corruption—for example, a revolving-door system where the same individual works for the pesticide company and then the regulator—you may believe that decisions about pesticide regulations do not protect the public. Those with this belief may protect themselves by consuming organic food, on which (synthetic) pesticides are not used. Some surveys suggest this is a major reason consumers in the United Kingdom and United States buy organic.

We, the authors, have confidence in the US and EU regulators, and believe pesticides in agriculture pose very few dangers to the safety of our food supply. In our view, the potential dangers of pesticides are outweighed by the benefits they provide in lowering the price of fruits and vegetables. However, we recognize that some readers will disagree, and will thus seek to protect themselves by purchasing organic food.

Is Organic Food Free of Pesticides?

No, organic food does contain pesticide residues. Synthetic pesticides are found on around 25 percent of organic fruits and vegetables. Such pesticides are not allowed under organic certification standards, suggesting that not all farmers are following the rules (note that conventional farmers sometimes deceive, too, as residues from banned pesticides are sometimes found on food). Still, the residues occur in much smaller amounts than in conventional food. When organic food is said to contain less pesticide residue, that claim ignores the "natural" pesticides organic producers are allowed to use. These are chemicals, biological agents, and minerals found in nature that do not need to be transformed using advanced chemistry and big factories. Rotenone is acquired from the roots of certain plants, and can cause neurological disorders. *Bacillus thuringiensis* is a bacteria found in the soil. Copper and sulfur products are minerals, and are both toxic at high levels. All of these are applied to crops to protect them from pests, and all can pose considerable health harms if used recklessly.

How dangerous are these organic pesticides, and do they make organic food less safe to eat than conventional food? First, it should be noted that organic farmers in most of the developed world can only use government-approved organic pesticides, and these are approved because they are deemed to be safe. There are natural pesticides that are prohibited because of their toxicity, such as nicotine, lead, and arsenic. Those that are allowed are usually exempt from the maximum tolerance levels because they have low toxicity, are unlikely to be detectable in foods, or decompose quickly, thereby posing few health risks. Most organic pesticides must be approved by the EPA and are subject to the same safety standards, so pesticide residues in organic food pose no more danger than residues in conventional food.

The consensus is that, while organic food contains fewer synthetic pesticide residues, it does not seem to improve health—but neither is it worse for health. The National Academies of Sciences has determined that natural and synthetic pesticides are equally safe, and 85 percent of toxicologists disagree with the statement that organic/natural products are safer in regard to chemical exposure. In a comprehensive review of organic foods, researchers found that consumption of organic produce doesn't increase one's exposure to pesticides, but that farmers who apply the pesticides face the most risk. Perhaps we need to worry less about pesticides in our food and more about pesticide exposures in our farmworkers? That said, the EPA *does* account for farmworker exposure to pesticides (and even pesticides used in the home, including insect repellent).

In regards to organic food, one must make a personal judgment. There is no compelling reason to fear organic foods, but no overwhelming evidence to express confidence in their safety either. Most people probably have an intuitive opinion about which foods offer the best combination of safety and nutrition. Hopefully this chapter on pesticides has made that intuition better grounded in facts.

3

THE CHEMICAL FERTILIZER CONTROVERSY

What Is the Chemical Fertilizer Controversy?

Chemical fertilizer has an almost miraculous impact on crop yields—so miraculous, in fact, that some people mistake it for a divine miracle. Evangelical Protestants in Guatemala have recently been stealing adherents from Catholicism, while at the same time the widespread adoption of fertilizers created a remarkable rise in vegetable production. Instead of giving credit to modern fertilizers, Evangelical ministers have been claiming the boost in food production was God's reward for their conversion.

> [A Guatemalan citizen states] "God started moving in the whole community, through different miracles that happened—people were getting saved, even I heard miracles of people being raised from the dead, [curing alcoholism], families being restored, you didn't see all the drunks on the street anymore. It was a complete change." And according to Evangelical lore, not only did the character of the people...change, God blessed and healed the land so they could grow more vegetables, and big vegetables, and the records show the vegetables got bigger.
>
> A leading Guatemalan economist published an article about [a Guatemalan region] in which he points out that

> the mass conversion of Evangelism occurred around the same time as chemical fertilizers, new seeds, and new crops were being introduced to Guatemala.
>
> —Sean Cole, "Spiritual Warfare: Evangelical Protestants Convert Catholics," *The Story*, American Public Media, March 19, 2013.

Chemical fertilizer is something of a miracle, though one born of human ingenuity. Instead of fertilizing the land through manure, leaving land fallow, or planting cover crops, we reach deep into the earth for phosphorus and potassium, and up to the sky for nitrogen. Modern chemical fertilizers are not necessarily better than ancient sources of enrichment, as they provide only some of the nutrients plants need, but they are far less expensive.

Without chemical fertilizers we could feed only 60–70 percent of the current population, some researchers believe. So reliant are we on nitrogen fertilizer that of all the nitrogen in the muscles and organs of humans, almost half of it was created in a fertilizer factory. The Amazon basin was once thought a poor location for any agriculture besides peasant farming, but because of fertilizers it has become an agronomic superpower. Chemical fertilizers are a blessing, and most agricultural scientists agree that they are the most important source of yield increases in the last century. Anyone who uses them will be astounded at their impact on plants, so much so that it is understandable that some Guatemalans can mistake these yield enhancements for a divine gift.

What is happening in Guatemala is simply a continuation of the Green Revolution—a revolution not of politics but of agriculture. The hero of the revolution was Norman Borlaug, who some say is a contender for the greatest American of the twentieth century. Developing new varieties of crops for the developing world in the 1950s and 1960s, and then teaching farmers how to raise them with modern fertilizers, Borlaug's movement increased the world production of food calories from 2,063 to 2,798 per person. Huge famines had been predicted in the second half of the twentieth century, but thanks to Borlaug the only famines were caused by politics (as in China). So influential was he in helping to feed a growing population that he was awarded the Nobel Peace Price in 1970.

> If they lived just one month amid the misery of the developing world, as I have for 50 years, they'd be crying out for tractors and fertilizer and irrigation canals and be outraged that fashionable elitists in wealthy nations were trying to deny them these things.
> —Norman Borlaug, responding to those who criticized him for espousing the use of modern agricultural technologies, quoted in Gregg Easterbrook, "The Man Who Defused the 'Population Bomb,'" *Wall Street Journal*, September 16, 2009, A27.

From the modern world's perspective, the positive developments in Guatemalan agriculture are no surprise. What is surprising is that it took so long for the Green Revolution to reach Guatemala.

Is the miracle of chemical fertilizers too good to be true? Some think so. While they do not deny the ability of chemical fertilizers to improve agricultural productivity in the short run, they argue the long-run view is not so optimistic. Critics also argue that chemical fertilizers lead to pollution and encourage the growth of large corporations, and for some, large corporations are themselves a problem. These are the controversies we will now explore.

Do Chemical Fertilizers Enhance Soil Fertility?

The obvious answer seems to be yes. Why else would farmers use them? As farmers harvest wheat, corn, and other crops, they are taking from the land all the elements within that crop—all the nitrogen, phosphorus, potassium, carbon, water, and other minerals. Unless the crop is used in a small community where all the excrement of animals and humans are returned to the soil, a field immediately becomes less fertile after harvest, so farmers return nutrients to the soil before planting the next crop.

Most farmers in the modern world do not return all the nutrients they take from the land. Chemical fertilizers typically consist of only nitrogen (N), phosphorus (P), and potassium (K). Farmers will periodically apply lime to restore a proper pH balance in the soil. However, they do not apply many micronutrients because it would not increase yields. Plants need very few micronutrients in any one year and ample amounts are already present in most soils. A number of experimental fields at universities have raised crops for decades (in some cases over a century), and yields continue to rise even when fields receive only N, P, and K (and occasionally the micronutrients sodium and magnesium).

Many farmers to-day are anxious because they can no longer make good farmyard manure, but must rely more on artificials, and grow cereals more often. Will they injure the soil? The Broadbalk results show that, apart from disease, the yield of wheat can be kept up indefinitely by proper artificials.
—Sir John Russell, director of Rothamsted Station, 1943, quoted in Philip Conford, *The Origins of the Organic Movement* (Edinburgh: Floris Books, 2001).

Has chemical fertilizer enhanced fertility in the last one hundred years? Absolutely, but with one qualification. Most of the yield gains witnessed over the past seventy years did not derive from chemical fertilizer alone, but from new crop varieties, machinery, and pesticides as well. In fact, new grain varieties were sometimes created in response to the availability of chemical fertilizers, as researchers sought varieties that would consume N, P, and K in greater quantities.

Although the impacts of chemical fertilizer are almost miraculous, it is their cost—not their direct impact on yields—that is so astounding. Crop yields can be just as high using organic fertilizers like compost and manure, but the availability of organic manure is constrained by animal populations, our willingness to eat food fertilized with human manure, our eagerness to use products that can be composted (like potato chip bags made out of corn), and the high cost of transporting and applying manure and compost. Industrial processes of making chemical fertilizers have been around for more than a century, but technologies have improved considerably. The price of nitrogen fertilizer, for example, has fallen 90 percent since 1900.

Is Chemical Fertilizer Enough to Ensure Soil Fertility Forever?

No, for several reasons. First, chemical fertilizers are created using nonrenewable resources, and it is not known if they are economical using renewable sources. Nitrogen depends on natural gas, and phosphorus and potassium are mined. However, almost everything that takes place on a farm is based on nonrenewable resources, including the tractor driven by an organic farmer. Even many Amish farmers operate gasoline-powered generators. Second, plants require more than N, P, and K, and eventually the soil will be depleted of its micronutrients, like boron and copper. Third, these chemical fertilizers alter the soil's pH, which may prevent the plant from being able to consume the nutrients. Fourth, some people have an alternative take on the definition of soil fertility, and may deem a field infertile whenever it lacks high amounts of organic matter, even if that field provides high yields (but this view is rarely held by agricultural scientists).

Micronutrient depletion is of particular interest because it raises the question of how long chemical fertilizers can be used as the sole nutrient source. It should first be noted that using organic fertilizer like manure can result in excessive levels of these micronutrients and other trace elements. Repeatedly applying animal manure to cropland can lead to such high levels of copper and zinc that it becomes toxic to plants. Applying too much of the wrong kind of manure can make a field less fertile. This does not happen quickly, though. Livestock manure consistently applied for ten or even twenty years at reasonable levels does not lead to excessive levels of trace elements.

Most fields receiving only chemical fertilizers continue to increase in productivity, so these micronutrients do not appear to threaten food production yet. One researcher found that for one particular region, the soil's store of nitrogen could feed a crop for twenty years, while its store of micronutrients could feed a crop for thousands of years. Fields receive micronutrients from atmospheric deposition as well. This helps explain why farmers can harvest more and more crops over decades or even centuries without depleting the soil of micronutrients. Still, we know many of these micronutrients will one day need replacing, and some sooner than others, like the impending need for copper in the SanJaoquin Valley. These micronutrients pose so few problems that it is very difficult to find information about their decline over time, making it problematic to predict when most soils will require something other than N, P, and K.

Sometimes a micronutrient problem exists even though it is prevalent in the soil. This is because not all nutrients are available to plants. North Dakota soils have large amounts of iron, but much of it is unavailable because the soil's pH is too high. The solution is not to apply more iron but to lower the soil's pH, or to apply a chelate, which helps transport iron from the soil onto the surface of plant roots.

We do have an answer for what will happen when a soil becomes deprived of micronutrients, because in a few areas it has already happened. In the last twenty years some wheat farmers in the state of Washington noticed that applying more nitrogen did not increase yields, which suggested the wheat's growth was limited by the absence of two other micronutrients: chloride and sulfur. In response, fertilizer companies developed a market to profit off this need, giving farmers access to inexpensive chloride and sulfur, and thereby restoring high yields. When some other micronutrient becomes scarce in the soil, fertilizer companies and farmers will respond in the same way. They have already done so for zinc, borax, manganese, copper, and iron. In fact, in areas like Oklahoma, where micronutrients are rarely a problem, there are salespeople trying to sell fertilizer supplements. However, agronomists typically find they do not increase yields. Just think how many salespeople will be selling copper, iron, and boron when it is actually needed! Clearly, the fertilizer debate concerns more than the scientific principles of agriculture. It is also about whether markets will respond to fertility problems before it is too late. Given that these markets already exist, there seems little to fear.

Any agronomist will tell you that the soil is a complex, living ecosystem. It can't be described solely in terms of nutrients. For example, in order for soil nutrients to be available to the plant the soil must have a certain pH level. An ideal soil is "alive" in that it contains a gallery of worms, insects, bacteria, and fungi. Worms dig tunnels, helping plow the soil and drain water. Some insects eat the pests that destroy crops. Certain types of fungi live on plant roots and grab carbon from the air, storing it in the soil for long periods. A living soil can help plants in other ways that are difficult to verify or observe. For instance, some plant diseases are caused by a "dead" soil that allows disease to run rampant. How is a farmer to know what is really to blame: the disease itself or a dead soil?

There are fungi living on plant roots that help the plants communicate with one another, warning each other of an approaching pest, and allowing the plants to erect defensive measures. A variety of other beneficial microorganisms live in a healthy soil. The bacteria *Paenibacillus* found in California soils prevents tomatoes from being contaminated with salmonella, and some microbes help plant roots uptake phosphorus from the soil. Chemical fertilizers—especially anhydrous ammonia—can kill these microorganisms, although research generally finds their impact on microbial communities to be small. However, in cases where the impacts are larger, there are ways to respond without giving up chemical fertilizer. The absence of worms can be compensated for by aerating the soil mechanically. Companies like Terra-One sell fungi and other microorganisms to inject back into the soil. Gardeners have been buying such microorganisms for decades.

Healthy soils are also high in organic matter (mostly the decaying residues of previous crops), which help to prevent soil erosion and increase the soil's capacity for retaining moisture. Farmers can also return organic matter to the soil by adopting no-till methods, where the soil is not penetrated by a plow. They have also employed alternatives to chemical fertilizer, like livestock manure. Planting legumes between crops (what is called a "cover" crop) not only increases nitrogen in the soil (legumes are great at doing this) but since the legume is not harvested it remains on the field and becomes assimilated into the soil as organic matter. One American farmer adopting these innovations has even put a monetary value on this organic matter: $3,775 per acre.

Finally there is the pH issue. Through the application of chemical fertilizers (and other practices) soils are becoming more acidic over time, and once the pH balance is too low, plant yields fall. The solution is simple: put something on the ground to raise the pH. Organic fertilizers can sometimes achieve this, but the most common solution—a solution centuries old—is to apply lime. Some regions like the state of Washington have no local source of lime and the transportation costs are too high to import it, so farmers have watched their pH levels fall with no immediate solution. There is talk of a company creating something called "liquid lime" that might one day be economically feasible. If not, these farmers may have to switch to organic fertilizers to restore their field's productivity. Chances are, though, that some company will develop a solution for these farmers, something produced in large factories using advanced chemistry and owned by a large corporation.

For now, farmers keep harvesting more from their fields while relying almost exclusively on chemical fertilizer, and many are optimistic that they can continue to do so for a long time. Some do not share this optimism, probably fearing that the soil will become infertile very quickly, before they have time to adjust. This is why they champion the organic food movement, which asks us to think about soil supplements now, because a healthy soil must be nurtured constantly—not reacted to by necessity. Organic advocates may not believe that the private sector will quickly and cheaply provide soil supplements when they are needed. If chemical fertilizers are provided largely by corporations (they are) and one distrusts corporations, then skeptics may have little confidence in the private sector to save them in the future.

Do Chemical Fertilizers Reduce the Nutrient Content of Food?

Consumers hear conflicting reports about foods' nutrient content. Food writer Michael Pollan has claimed that fresh produce is 40 percent less nutritious today than in 1950, and this was written in a context that suggested modern technologies like chemical fertilizer are the problem. Writers for the *Scientific American* have made similar statements. The main culprit in this disturbing nutritional trend is soil depletion of nutrients, they say. If chemical fertilizers do not return micronutrients to the soil, it seems logical that food will also possess fewer micronutrients.

For an illustration of how a lack of micronutrients can impact health, consider iodine. Most readers will not experience a lack of iodine because it is added to table salt. However, for reasons of taste or religion, some people use only kosher or sea salt. This will not pose a problem so long as they eat vegetables from soils sufficient in iodine, but in some regions this is not the case.

There are many challenges when comparing food nutrients today with food nutrients in the past, but most evidence suggests that fruits and vegetables today are less nutrient-dense. The decrease is not alarming, though, as the USDA has shown that the nutrient composition of food hasn't changed much in the last century. Moreover, in cases where the nutrient composition fell, soil quality was not necessarily the problem. Different nutrient measurement methods, alternative ways of handling food, and new crop varieties are also a factor. Consumer preferences matter also. Although wild crab apples contain more phytonutrients than modern farmed apples, crab apples are barely edible. What good is a nutrient-dense food if no one eats it? Also, it should be noted that there are some cases where farmers and breeders are actively trying to increase the nutrient content of food.

Are chemical fertilizers to blame for the decline in nutrient density? Research suggests the blame mostly belongs to new crop varieties. With the rise of chemical fertilizer also came higher-yielding crop varieties. These more efficient varieties of grains, fruits, and vegetables are subject to the genetic dilution effect, a concept describing the trade-off between yield and nutrition. These improved varieties of plants achieve a higher yield in two ways: by taking more nutrients from the earth and by packing less nutrients per unit of food. Thus, the new crop varieties are probably the major source of nutrient loss in foods today. This was illustrated nicely on the Broadbalk fields in England, where experimental plots have been maintained since 1843. Between 1843 and the 1960s the concentration of micronutrients in the wheat harvested remained steady, but then began falling, giving the impression that the soil was running out of zinc, iron, copper, and magnesium. However, the amount of these micronutrients in the soil had remained steady or increased, leading researchers to conclude it was the choice of wheat varieties planted that reduced the nutrient content of the wheat, not soil deficiencies.

Of course, the nutrient content of food is less important than access to total nutrients. What good is a nutrient-dense food supply if there is little of it to go around? Even if the nutrient content of food has fallen, agriculture has become increasingly productive, suggesting the total per capita nutrients in the food supply might be rising. Indeed, that seems to be the case. Figure 3.1 looks at how total access to nutrients in the US food supply has changed in the last century, and what it shows is true using a variety of measures, like food energy, protein, vitamins, and minerals. The results are clear and striking: our access to nutrients has risen over time in every measure except potassium, which declined only slightly. Increases in magnesium, vitamin B12, and selenium were unremarkable, but all things considered, if soils are becoming less fertile, it is not manifested in the total amount of nutrients available to US citizens. The same is likely to be true in western Europe.

Greater nutrient availability doesn't necessarily mean greater nutrient consumption, if patterns of food waste are changing at the same time more nutrients are produced. Studies of nutrient consumption in the United Kingdom find that per-person intake of some micronutrients like magnesium, iron, zinc, and copper has fallen over time, and in some cases is insufficient for a person's daily nutrient needs. While there are nutrient deficiencies in the United States also, they have not changed much since 1999. The purpose of including figure 3.1 is not to imply that there are no micronutrient deficiencies in the United States, but that the persistent use of chemical fertilizers does not seem to pose a micronutrient problem.

Another way to inquire whether chemical fertilizers affect the nutrient content of food is to compare nonorganic food to organic food. Most of the time nonorganic food is raised using mostly chemical fertilizers, whereas organic food must use other sources. Many scientific comparisons have been made,

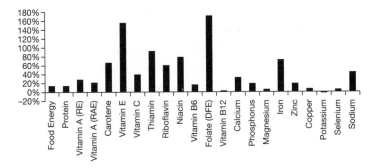

Figure 3.1 Percent Change in Nutrients Contained in Per Capita US Food Supply Compared to a Century Prior

Notes: Numbers are calculated as the percentage change in average, per day, per capita nutrient supply over the period 1997–2006 relative to the average over the period 1909–1918.

Source: Economic Research Service, *Nutrients (Food Energy, Nutrients, and Dietary Components)* (dataset), US Department of Agriculture, accessed July 30, 2013, at http://www.ers.usda.gov/data-products/food-availability-(per-capita)-data-system.aspx#26715.

but the overall results suggest organic food is slightly superior or equivalent to conventional food in terms of nutrition. When grocery stores in the United Kingdom tried to market organic food as being more nutritious, they were ordered by the government to stop, because it was considered false advertising. The stores could not refute the government's accusation, so they ceased advertising organic food as nutritionally superior. Organic food does have fewer pesticide residues, but we defer this issue to the chapter on pesticides.

Does Chemical Fertilizer Cause Too Much Water Pollution?

There are two ways of measuring food production in a region. One is to look at the actual amount of food produced. Another is to measure the amount of water pollution. Both the United States and China, which have greatly increased their food production in the last fifty years, have seen a decline in water quality. Over 64 percent of US lakes are impaired in that they cannot be used for fishing and swimming. The numbers for

US rivers and estuaries are 44 percent and 30 percent, respectively. More than half of lakes in China suffer from too much fertilizer. Groundwater can be contaminated by excess nitrogen also, and can cause blue baby syndrome and other health problems. Even Austrian springs considered by the Catholic Church to be holy water are contaminated with enough nitrogen to cause illness. The damage to waters are thus caused by the success of modern agriculture. The challenge now is to restore these waters while maintaining an adequate food supply.

Not all the fertilizer applied to a field will be consumed by the plant. Some will leave the field as surface runoff or subsurface leaching. Though that excess fertilizer will not feed the crop, it will feed something. It may fertilize plants growing on the side of a field, trees downhill from the field, or bacteria and algae in rivers. If enough fertilizer reaches surface waters, it may cause explosions in bacteria and algae populations, and as the populations expand they consume more oxygen from the water. Eventually the water might reach eutrophication, where oxygen is so scarce that no aquatic life can exist. The water becomes cloudy, and using the water for drinking may require expensive treatment. In the Gulf of Mexico there is a "dead zone" of about 3,100 square miles, so this is no minor problem.

Restoring water quality first requires us to understand all the sources of eutrophication, since chemical fertilizers are not the only one. Livestock manure is responsible for most of the pollution in the Chesapeake Bay and the Illinois River. In other areas lawn fertilizers cause the most damage. Phosphorus in dishwashing detergents was a major pollutant in the US Great Lakes, which is why most detergents now say, "Contains no phosphorus."

It is simply hard to apply fertilizer of any kind and not experience some runoff. Roughly half of all applied nitrogen fertilizer will not be consumed by the crop, which means it fertilizes other plants or enters waters. When agronomists make recommendations on how much nitrogen to apply, they often just multiply the amount of nitrogen the crop needs by two. For phosphorus around 30 percent of the amount applied isn't consumed by the crop.

It is becoming evident that farmers can apply less fertilizer with little sacrifice in yield, especially in some developing countries. Farmers apply excessive fertilizer for a number of reasons. In China it has to do with memories of famines in the Mao era, in India it is related to the large fertilizer subsidies, and in Japan it is partly because so many rice producers are only part-time farmers who do not have time to perfect their management skills.

One obvious solution is to remove fertilizer subsidies where they exist and perhaps even tax fertilizers, but this is difficult in poorer countries with food security issues. A tax is more feasible in richer countries. In 2013 the California Water Resources Control Board actually recommended placing fees on fertilizer use, in order to help offset the cost of treating drinking water polluted with excess nitrogen. A 1992 fertilizer tax in Sweden reduced use by 15–20 percent, and if farmers were indeed applying too much before the tax, the sacrifice in yield may have been minimal. With the support of both farmers and environmentalists, the state of Illinois instituted a fertilizer tax not to discourage its use, but to fund research on its environmental impact and ways to reduce runoff.

New technologies in precision agriculture can vary the fertilizer application rate across a field, probably reducing excess fertilizer in the process. Called "filter strips," a buffer of unfertilized permanent grass can be placed between the crop and the edge of a field to catch fertilizer runoff. Studies have shown that filter strips can eliminate over half of all fertilizer runoff, as well as reduce soil erosion. However, they can be costly because they reduce the amount of land in crops and require maintenance, so they are sometimes subsidized by agencies like the USDA or individual states like Iowa. Minnesota has a ban on crops within fifty feet of a stream to ensure a vegetative buffer between the crop and surface waters. Through a variety of activities—some subsidized and some undertaken by farmers at their own expense—nitrogen and phosphorus runoff from fields in the United States has fallen around 21 percent and 52 percent, respectively.

In some areas livestock manure is the biggest problem, but this can be solved with better manure management regulations. When a farmer is deciding how much manure to apply on each acre, she was previously required to match the nitrogen content of the manure with the nitrogen needs of the crop. However, because regulations allowed her to ignore phosphorus, the result was phosphorus runoff. Newer regulations now require the farmer to match the nitrogen *and* phosphorus content of the manure to the needs of the plants. Further, farmers can inject manure into the soil instead of spraying it, which should reduce total nutrient runoff. Some new manure management regulations could backfire, so they must be written with care. For instance, the new phosphorus regulations might reduce phosphorus runoff, but at the expense of greater nitrogen runoff.

Could we reduce fertilizer runoff by switching to organic production? Experiments from Michigan fields suggest yes. They find that nitrogen runoff is lower in organic systems compared to systems using chemical fertilizer, even when reduced levels of chemical fertilizers are applied. One would suspect that the organic yields were lower, but even if that were the case, the nitrogen runoff was still lower on the organic fields for each unit harvested. This does not mean organic systems always lead to less pollution, for in these experiments nitrogen was supplied to organic fields not by the applications of livestock manure but through nitrogen-fixing legumes. The use of no-till systems in the experiments were found to reduce nitrogen runoff, but other studies have found that no-till can actually increase phosphorus runoff (partly because the phosphorus is spread on the field surface instead of being tilled into the soil). The lesson is that farmers can take actions to reduce water pollution, but there is no silver bullet that works for every farmer in every location. The choice of whether organic or nonorganic, till or no-till methods should be used depends on the source of organic manure and whether nitrogen or phosphorus runoff is a greater problem in the region, among other things.

Everywhere that chemical fertilizer and livestock manure are applied to land there are some water pollution issues. Sometimes they are hardly recognizable and sometimes they result in eutrophication. The good news is that scientists can easily detect water quality problems, and solutions for mitigating the problems are known. The problem is solvable, but the motivation to solve it isn't always strong.

For example, on the one hand, the Chesapeake Bay is a relatively small area, and its runoff problems are caused by local sources. Citizens seem ready and willing to pass the kinds of regulations that should eventually restore its quality. They are willing to pay the cost because they will receive the benefit. On the other hand, there seems to exist a fatalistic attitude towards the dead zone in the Gulf of Mexico. The zone is caused in part by every farm within the Mississippi basin, which includes Louisiana, Nebraska, Iowa, Ohio, and even Montana, just to name a few contributors. What is the likelihood that Montana wheat farmers and Iowa hog farmers will cooperate to reduce nutrient runoff into the Gulf of Mexico—a gulf very few of them will visit? For these reasons, small water sources affected by only local communities are more likely to solve the fertilizer problem. In areas like the Gulf of Mexico, however, there is little prospect for a solution other than strict federal regulations, and that solution has few supporters. There is little doubt that a very large fertilizer tax would reduce runoff, but policies that hike food prices are not very popular with citizens, and therefore, are not popular with politicians either.

4

THE CARBON FOOTPRINT CONTROVERSY

What Is the Carbon Footprint Controversy?

Nearly all humans consume meat, dairy, and egg products in some form. In recent years the environmental movement has touted the necessity of reducing one's "carbon footprint." Can we reduce our footprint without changing our diet? Much controversy surrounds that question. One very extreme view on the political left is below.

> But when it comes to bad for the environment, nothing—literally—compares with eating meat. The business of raising animals for food causes about 40 percent more global warming than all cars, trucks, and planes combined. If you care about the planet, it's actually better to eat a salad in a Hummer than a cheeseburger in a Prius.
> —Bill Maher, host of HBO talk show *Real Time with Bill Maher*, writing in the Huffington Post in 2009, accessed April 25, 2013, at http://www.huffingtonpost.com/bill-maher/new-rule-a-hole-in-one-sh_b_259281.html.

The last decade has seen a movement advocating a vegan diet in order to reduce carbon emissions, and in some respects the argument is logical. After all, it takes about 3.388 pounds of corn (and many other inputs) to produce a single pound of retail beef, making meat seem relatively inefficient in comparison to grains, thus leading to a larger carbon footprint. So common is this notion that some schools encourage "Meatless Mondays" for the sake of the environment. The Meatless Monday movement has even been adopted by the Norwegian military. Moreover, there is scientific research showing that vegan (and vegetarian) diets do result in a smaller carbon footprint.

> When dealing with issues as big as global warming ... it's easy to feel helpless, like there's little we can do to make a difference.... But the small changes we make every day can have a tremendous impact. That's why this Meatless Monday resolution is important. Together we can better our health, the animals and the environment, one plate at a time.
> —Los Angeles councilmember Ed Reyes, coauthor of a Meatless Monday resolution in 2012.

However, equally prestigious research shows that vegan diets can result in a higher carbon footprint. How can this be? One reason is that some carbon footprint estimates are wrong, or rather, interpreted incorrectly. The idea of livestock production being a large carbon emitter began with a report by the United Nations (UN) suggesting that livestock contributes 18 percent of the world's carbon footprint, more than the transportation sector, thus giving Bill Maher reason to point the blame at burgers instead of Hummers.

It turns out that this 18 percent estimate is fraught with errors, or at least, doesn't represent conditions in the United States (perhaps the results are valid for the world). For instance, the UN did not account for the carbon emissions involved in making the inputs used in the transportation sector, but they did for livestock. This would be like saying the production of tires requires no carbon emissions but the production of corn does. Also, that 18 percent makes a number of contestable assumptions, especially regarding how land use changes as livestock production rises. Finally, the study was meant to be an estimate for world emissions, whereas people like Maher were interpreting it as relevant to US emissions. Correcting these mistakes for the United States shows that livestock is responsible for only about 3 percent of Americans' carbon footprint, whereas transportation counts for 26 percent. All of agriculture impacts only 6–8 percent of the American footprint (note that these numbers only refer to carbon emissions that are the result of human activity, not emissions occurring naturally).

Much happens with food between farm and fork, and carbon is emitted there also. Only 20 percent of the cost of food reflects farm activity. The other 80 percent reflects the cost of labor, energy, machinery, and other activities at the food-processing and retail levels. A food item might generate few carbon emissions at the farm but large emissions at the processing level, so a vegan diet might not be light on carbon if it undergoes considerable processing. One must also account for the amount of food eaten, and the extent to which people want to eat it. Although 3.388 pounds of corn may be needed to produce one pound of retail beef, each pound of beef has more calories than a pound of corn, and beef and corn provide very different eating experiences. To determine the relationship between diet and carbon footprints the best studies examine the actual foods that vegans, vegetarians, and omnivores consume. Because these studies do not agree, a controversy exists.

The crux of the debate has moved beyond the question of vegetable versus meat diets, and instead attempts to find out the types of food that provide the most satisfaction with the smallest carbon footprint. This chapter will compare organic to nonorganic food, beef to chicken, and grass-fed to corn-fed beef, and the relationship between carbon emissions and the price of food.

The reader might wonder why we haven't mentioned the controversy of global warming itself, given the torrid rhetoric between global warming denialists and alarmists. Because we are not climate scientists we will say very little about how agricultural activities affect future temperatures. We do assert two statements as facts, though. One is that gases like carbon dioxide and methane are greenhouse gases that retain heat. After all, Venus is much hotter than Mercury despite being farther from the sun, simply because the atmosphere of Venus contains greenhouse gases and Mercury's does not. The second fact is that, because some activities take greenhouse gases from the earth's crust and eject it into the atmosphere, there is some probability that the use of fossil fuels will have a noticeable effect on the climate within the next one hundred years. We make no assertion as to what that probability is, only that it is greater than zero. This chapter focuses on how food affects emissions of greenhouse gases, not on how those emissions affect the climate.

Throughout this chapter we will be referring to "carbon" when we are really concerned about all greenhouse gases, and so "carbon" really refers to carbon dioxide-equivalent emissions—denoted CO_2e. One ton of methane results in twenty-one times the warming of 1 ton of carbon dioxide, so if a ton of methane is emitted, we instead say that 21 tons of "carbon" emissions (20 tons of CO_2e) take place. To repeat, when we say 1 ton of carbon, we are always referring to 1 ton of CO_2e.

How Does Agriculture Contribute to Greenhouse Gas Emissions?

In the 1920s, when Russian novelist Yevgeny Zamyatin wrote his science fiction dystopia *We*, he imagined food being produced directly from petroleum. His vision was prescient. Plants may acquire their energy from the sun, but farmers acquire it from fossil fuels. Obviously, tractor fuel is based on petroleum, but so is nitrogen fertilizer. Other fertilizers like phosphorus and potassium require mining, using the brute force of oil-powered machinery to dig deep into the earth. Pesticides require fossil fuels, so do irrigation equipment, milking machines, and so on. Even if organic farmers use manure as fertilizer, they still rely upon fossil fuels, and not just to power machinery. Often they acquire this manure from livestock on nonorganic farms, and that livestock was fed forage and grain fertilized by chemicals that were created using fossil fuels.

> I refer to the great Two Hundred Years' War, the war between the city and the land. Probably on account of religious prejudices, the primitive peasants stubbornly held onto their "bread." In the thirty-fifth year before the foundation of the One State our contemporary petroleum food was invented.
> —Yevgeny Zamyatin, *We* (New York: Dutton, 1924).

Since most of the new carbon ejected into the atmosphere comes from fossil fuels, the carbon footprint of food can be reduced by using less oil, coal, and natural gas. Because these fuels are expensive, farmers and food manufacturers have always been in the game of conserving energy, and have done so mostly by increasing their productivity. As technology has evolved over time to reduce energy costs, the carbon footprint of food has likewise fallen.

There is more to a carbon footprint than energy use. One reason beef has a larger footprint than pork or chicken is because cattle are ruminants, and ruminants expel carbon as they burp—and they burp about once a minute. A wheat farmer who plows his soil might emit more carbon than a wheat farmer using no-till methods, as breaking and turning the soil releases carbon into the atmosphere.

The carbon footprint of food even depends on consumer behavior. It takes more gasoline to drive to the grocery store *and* to the farmers market, and consequently, more carbon emissions. The relationship between food and carbon does not depend solely on how food is produced, but also how it is purchased.

Does Organic Food Have a Lower Carbon Footprint?

It depends. Carbon emissions are influenced by the type of nitrogen fertilizer a farm uses. Organic advocates are quick to point out that they do not use chemical nitrogen fertilizer, and thereby avoid one large source of carbon emissions. Many organic farmers acquire their fertilizer from livestock manure, though, and many of these animals were fed forage that was fertilized with chemical fertilizers. If this is the norm, then organic farmers are reliant on chemical fertilizers but are using it in a very inefficient manner that might increase the carbon footprint of their food. Manure and human compost—two sources of organic manure—both emit carbon as they are stored, and these emission rates are rather large.

An advantage of organic fertilizer is its ability to sequester carbon in the soil. Fertilizing a field with manure, compost, or cover crops doesn't just add the three key nutrients of nitrogen, phosphorus, and potassium, but also increases the carbon content of the soil. Such fertilizers are not strictly reserved for organic production, though, and there are many farms raising conventional crops using livestock manure. Still, most conventional farms do not use organic fertilizer, while all organic farms do. If a heavily farmed field low in organic matter is converted to organic production, carbon will be extracted from the atmosphere and stored in the soil, thereby reducing the carbon footprint of organic food. The rate at which carbon is sequestered in the soil is highly uncertain, though, and causes considerable uncertainty in carbon footprint measurements.

Although organic producers may argue otherwise, organic farming is less productive (this is discussed in detail in another chapter). Organic farmers, by definition, have a smaller number of farming options to choose from than conventional farmers. In fact, if organic really were more productive, there would be nothing stopping conventional farmers from using organic methods, but the fact that conventional farmers choose other technologies suggests organic methods are less productive.

Higher productivity means it takes fewer inputs to produce any given unit of output, and fewer inputs typically mean less carbon emissions per unit. However, an "input" is not a single thing, and even if a farmer spends less money on all inputs, that doesn't mean the carbon emissions from those inputs will fall, if the portfolio of inputs changes in favor of high carbon emitters. For these and other reasons, the question of whether organic food has a smaller carbon footprint is an empirical question, requiring data for an answer.

The data do not crown either organic or conventional as king. Sometimes organic food has less emissions and sometimes it does not. One study comparing organic and conventional production of twelve crops (blueberries, two kinds of apples, two kinds of wine grapes, raisins, strawberries, alfalfa for hay, almonds, walnuts, broccoli, and lettuce) found that conventional production usually had a smaller footprint, assuming the land had been used in the same way for many years. However, if the organic farm is rather new, and its fields were heavily plowed in the past, the soil may be sequestering enough carbon to make organic produce the low-carbon emitter.

Similar results were found for hogs. Conventional hog production appears to emit less carbon when soil sequestration of carbon is ignored, but, when it is accounted for, organic pork systems might or might not emit less carbon. Beef will be discussed in its own section, and it will be shown that unless carbon sequestration of pasture is much higher than current measurements, organic beef results in a larger carbon footprint.

There is insufficient evidence available to state that organic agriculture overall would have less of an environmental impact than conventional agriculture. In particular, from the data we have identified, organic agriculture poses its own environmental problems in the production of some foods, either in terms of nutrient release to water or in terms of climate-change burdens. There is no clear-cut answer to the question: which "trolley" has a lower environmental impact—the organic one or the conventional one?

—C. Foster, K. Green, M. Bleda, P. Dewick, B. Evans, A. Flynn, and J. Mylan, *Environmental Impacts of Food Production and Consumption: A Report to the Department for Environment, Food and Rural Affairs* (Manchester: Manchester Business School; London: Defra, 2006), 14.

A French study compared fifteen different foods and found organic versions of the food to have a lower footprint in only five cases. There appears to be no noticeable increase or decline in carbon emissions from switching from conventional to organic milk production. When comparing meat products in the United Kingdom, organic production resulted in lower emissions for sheep and pork, but higher emissions for beef and poultry.

The frustrating thing about evaluating the carbon footprint of organic and nonorganic food is the blatant biases of sources. For instance, the Environmental Working Group published a figure showing the carbon footprint of various foods, including fruits, vegetables, meat, dairy, and eggs. They communicate the idea that chicken meat is the best meat in terms of carbon, but then they added the qualification that this chicken should be organic, pasture-raised, and/or antibiotic-free whenever available. This suggests that organic chicken emits less carbon than conventional chicken, but the report cited does not study any organic foods, and certainly doesn't compare organic to nonorganic food. It is clear that this source took the results from (what appears to us to be) a credible study and then added some of their own beliefs (which cannot be defended by data), making it seem as if that credible study deemed organic foods to be environmentally friendly.

The agricultural industry is also to blame for misleading readers. The agribusiness newspaper *Feedstuffs* has published a number of articles making it seem that organic food always has a higher carbon footprint, yet research contradicts such blanket conclusions.

Often it seems groups decide first which foods emit less carbon and then seek out the data defending their choice, rather than letting the data determine their conclusions. This way of thinking detaches a controversy from evidence, making resolution impossible.

How Do Animal- and Plant-Based Foods Compare?

Two studies have been conducted to answer this question by studying the actual foods vegans, vegetarians, and omnivores tend to eat. One study took place in the United Kingdom and one in France, and they yield conflicting results. The UK study found that the foods vegetarians and vegans eat result in less carbon emissions. The French study found that, on a per calorie basis, fruits and vegetables had a carbon footprint similar to animal-based foods, except meat produced from ruminants like cows and sheep, whose emissions were larger. So long as beef is a major part of the omnivorous diet, it probably has a higher carbon footprint than vegan diets. While animal-based foods emitted more carbon on a per pound basis, one generally needs more pounds of fruits and vegetables to provide the same number of calories, and so they concluded that vegan and vegetarian diets do not have lower carbon footprints.

Suppose the UK study is right and vegan diets do emit less carbon. Vegan meals are also less expensive, according to this study, and if those savings are spent on things that are high emitters of carbon (like a plane flight) they may negate any emission reductions attributable to the vegan diet directly. The point isn't that vegan diets are not better for the planet, but that one should consider one's entire carbon footprint, not just that related to food.

The impact of one's diet even depends on where one is located. Those located close to fertile land, where fruits and vegetables can be produced in ample quantities, may be able to lower their footprint by eating a vegan diet. On the other hand, an omnivorous diet with moderate amounts of meat emits the least carbon in areas with poor soil, where grass for livestock may be the only good use for the land. Why? Because efficient food production requires putting agricultural lands to their best use. It is best to grow tomatoes in Southern California and beef in Montana.

How Do Beef, Pork, Eggs, and Poultry Compare?

Hogs and birds in livestock production can convert feed to meat more efficiently than cattle. They also reproduce faster. Cattle grow and reproduce slower, and, being ruminants, emit greenhouse gases as they burp. For these reasons the carbon footprint for beef is about three times the size of pork and turkey, four times the size of chicken, and six times the size of eggs (all on a per-pound basis). The magnitudes differ across studies but there is little debate that beef has the largest footprint, followed by pork, then chicken, then eggs.

Should We Feed Cows Grass or Corn?

Readers may have seen the label "grass-fed" beef at specialty stores or farmers markets. This is a certified USDA label that beef producers can earn so long as they feed cattle only forage (grass and hay) throughout cattle's life and always provide them access to pasture during the growing season. The label may be slightly misleading, as all cattle spend a large part of their lives on pasture, even if they are not sold as grass-fed beef. The difference comes in the last four to six months of the cow's life, where most cattle enter a feedlot and are given a diet consisting mostly of grain (usually corn and soybean meal). Grass-fed beef do not enter a feedlot, but remain on pasture whenever grass is growing. It would be more transparent to call one type of beef "corn-finished" and the other "grass-finished" beef, but that is not the terminology that has evolved.

Consumers naturally want to know whether grass-fed or corn-fed beef has a smaller footprint, but there is no easy answer. The documentary *Carbon Nation* argues that one of the most effective means for reducing greenhouse gases in the atmosphere is to encourage permanent pastures for grazing. The film's director and producer, Peter Byck, passionately argues this can be accomplished by allowing cattle to remain on grass throughout their lives.

Plants, with the help of fungi that live on their roots, naturally capture carbon from the atmosphere and store it in the soil, but modern cropping methods tend to disrupt this soil and kill these fungi, releasing the carbon back into the air. Converting fields that formerly grew corn for the feedlot to pastures, *Carbon Nation* argues, can reduce our carbon emissions by 39 percent. While the film cautions that "now this is new science, to be sure," it follows by saying, "but the early numbers are encouraging."

Viewers of *Carbon Nation* or any interview with Peter Byck might be quickly convinced that by replacing their regular beef with grass-fed beef, they are doing their part for the planet without giving up the foods they love. If those same viewers then turned the channel from *Carbon Nation* to the talk-show *Stossel*, they might then see the animal scientist Jude Capper argue otherwise.

> [Grass-fed cattle] have a far lower efficiency.... The animals take twenty-three months to grow versus fifteen [for corn-fed cattle]. That's an extra eight months of feed,

of water, land use obviously, and an awful lot of waste. If we have a grass-fed animal compared to a corn-fed animal, that's like adding almost one car to the road for every single animal. That's a huge increase in carbon footprint.

—Jude Capper, "Why Grass-Fed Beef Is Worse for Environment," interview by John Stossel, *Stossel*, May 6, 2011, Fox Business Video.

Capper and her colleagues have published studies in the peer-reviewed scientific literature showing that grass-fed, organic, and most "natural" beef production systems have a higher carbon footprint than conventional beef. Feedlots simply produce beef more efficiently than forage systems. By using fewer inputs to produce a pound of beef, less fossil fuels are required to produce those inputs, and on a per-pound basis the carbon footprint for corn-fed beef is smaller. Moreover, it takes longer to "finish" grass-fed cattle, which means the cow has to live longer to produce the same amount of beef, and during that time the cow constantly expels methane. Capper reports that a conventional beef production system requires only 56 percent of the animals needed to produce the same amount of beef as a grass-fed system, only 25 percent of the water, 55 percent of the land, and 71 percent of the fossil fuels. Because it is less efficient, the carbon footprint for grass-fed beef is 68 percent higher than that of corn-fed beef, Capper calculates.

So, who is correct: Byck or Capper? It depends on the rate at which the extra pasture needed for producing grass-fed beef can sequester carbon. Although sequestration numbers are highly sensitive to environmental conditions, most of the measurements suggest Capper is correct and that grass-fed beef would actually increase the carbon footprint of a steak.

Byck is then correct only if his optimistic assumptions about the ability of soil to sequester carbon are correct. *Carbon Nation* is absolutely right that plant growth can sequester carbon in the soil, and the transition of land out of heavily plowed cropland into pasture for cattle could (again, under optimistic assumptions) deliver enough reductions to reverse Capper's findings. Once that transition is complete and the land reaches a new equilibrium of carbon in the soil, then carbon sequestration ceases and the smaller carbon footprint then belongs to corn-fed beef. This suggests that any advantage grass-fed beef may have is either unlikely or fleeting.

So for now, corn-fed beef seems to have a smaller impact on global warming, but the science on this issue is young. Measuring carbon footprints requires many assumptions that are difficult to verify, and small changes in those assumptions can have sizable impacts. Thus, other studies attempting to measure carbon footprints could arrive at different conclusions. Much more research on this topic is needed, including replication and extension of Capper's work.

How Can I Lower My Carbon Footprint from the Food I Eat?

Meatpacking in nineteenth-century Chicago was a brutal, dirty business. It was a time when companies could dump waste into rivers without criticism. Gustavus Swift erected large meatpacking plants in Chicago where livestock were slaughtered and then shipped to New England by rail. He did not care about the purity of the local rivers, yet his polished business skills contributed remarkably to reducing water pollution.

Swift didn't make money from his sales of beef. His profits were in the by-products from beef production, like fat turned into soap, hides processed into leather, guts into tennis racket strings, and hair into stuffed cushions. The few parts of the carcass that were discarded left the plant in sewer pipes and flowed into Bubbly Creek, which then flowed into the Chicago River. Any amount of fat, gut, or hair that escaped the plant and flowed into the creek was money lost to Swift, so he would wade into the water to watch what came out of the sewer. Any fat escaping the pipe was a clear sign that his factories operated inefficiently, so he would trace the source of the leakage and fix it. He despised pollution, not because he loved the environment, but because he loved money. By pursuing his own self-interest he reduced the amount of pollution entering the river.

Why tell this story? Because we often ignore the fact that people every day reduce pollution without any intention of doing so, simply by producing things efficiently. The road to hell is paved with good intentions, the saying goes, so perhaps the road to heaven is built on self-interest. Every business that operates more efficiently than its rival reduces the carbon footprint of that product. Most people assume organic food is good for the environment because its proponents boast of their green intentions. Sometimes this is indeed the case. Yet sometimes nonorganic food has a smaller carbon footprint because it is produced more efficiently. Like Gustavus Swift, conventional producers look for every source of waste, and when they fix that waste they are able to produce each unit using less energy. Less energy then translates into less carbon. When the beef industry started giving calves growth hormones to help them grow faster, producers were not trying to reduce the carbon footprint of beef, but that is exactly what they did.

If readers really care about the environment, then they will concentrate on the actual level of carbon emissions and not just rely on the stated intentions of the seller. What matters are outcomes, and if an industry makes no announcements of sincere concern for the environment but leaves behind a smaller carbon footprint, it provides a public good.

As a general rule, the more a product costs, the larger its carbon footprint. This isn't always the case, but adding value to a product or service usually requires more energy, and most of the time that energy is derived from fossil fuels, which have carbon footprints of their own. This means that any time a person wants a higher valued product sold at a higher price,

she will probably leave a larger carbon footprint. Think back
to the carbon footprint of various meats, where beef's footprint
is larger than chicken's. It is also true that the price of chicken
is considerably less. You don't have to gather research behind
each food item to lower your carbon footprint—prices already
reveal a wealth of information.

Consider this example. A mug of coffee emits 23 grams of
CO_2e (carbon dioxide equivalent). Add milk, and that milk
must be produced from cows that emit greenhouse gases,
eat corn that was produced using fertilizers made out of fossil fuels, and produce milk that must be transported by truck
and kept cold in refrigerators. With milk, the coffee now emits
55–74 g CO_2. Cappuccinos entail even more activities, where
the milk must be steamed, increasing the emissions to 236 g
CO_2e. At each step, as the product increases in value—and
price—by requiring more inputs, more greenhouse gases are
emitted.

One step to reducing your footprint is to purchase less costly
food. Instead of getting a cappuccino, just get a regular cup of
coffee. The regular coffee also saved you money, though, and
you want to make sure those savings are not spent in a way to
increase emissions, so you would then take those savings and
purchase carbon offsets (where you pay others to emit less carbon). The same goes for other foods. Instead of going vegan,
where you might begin eating at expensive vegan restaurants
whose footprint might be large, consider eating a more austere
meal occasionally. Instead of purchasing grass-fed beef, consider purchasing regular, less expensive meat. Then be careful not to use the savings on yourself, but buy carbon offsets
instead.

Purchasing carbon offsets is easy, even for an individual
with just a little money. A simple Google search will reveal
many organizations like CarbonFund.org, where individuals pay money to prevent deforestation, restore forests,
encourage renewable energy production, and retire carbon
permits. These are activities to which you can direct your

savings from an austere diet. In fact, the average person's carbon footprint is about 24 tons of carbon per year, and at the CarbonFund.org website one can offset 24 tons for only $240. This means that if you are truly concerned about global warming but want to continue eating the same types of food, and consume other goods in the same proportion as before, for only $240 a year a person can shrink her carbon footprint to zero!

This doesn't mean you should not pay attention to claims made about a product's footprint. Nor are we suggesting that you ignore a food's nutritional content, calories, or other environmental impacts in addition to global warming. What we are trying to stress is that, in addition to all the environmental and health claims associated with a food, you should also incorporate information on the product's price in guessing its carbon footprint.

These are guidelines you can extend to every good you buy, not just food, and the goal is not just to reduce carbon emissions but to achieve that reduction by giving up the goods you value the least. Beef may have a larger footprint than chicken, but for many it is the most delicious meat, and they would rather reduce their carbon emissions by giving up anything other than beef. Evidence suggests that maintaining attractive lawns by chemical applications and regular mowing can cause the emission of more carbon per acre than land devoted to corn. A reader may decide to spend a lot less money on lawn fertilizer and eat the same foods, or he may decide the opposite.

5

THE GMO CONTROVERSY

What Is the GMO Controversy?

Suppose you are a citizen of France, China, or Germany—three among ten nations that consume horse meat. You enter a store to purchase a few pounds of the meat, but for some reason are worried that the owner is actually selling you meat from a different type of animal. Finding the store manager, you ask him whether he can guarantee that the meat labeled "horse meat" really came from a horse. He laughs and says, "Well, I raise the animals myself, and I can guarantee that every animal was born from a horse!" He is mocking you, and though too embarrassed to argue, you are still suspicious, and after buying the meat you send it to a lab for testing.

When the results come back it turns out that the owner was deceiving you and telling the truth at the same time. The meat came from a mule, not a horse, but mules are the offspring of a male donkey and female horse. Is mule meat so different from horse meat that the store was engaging in false advertising, or is a mule close enough to a horse that the store's actions were acceptable? That depends not so much on laboratory tests but whether consumers believe horses and mules are basically the same type of animal. Likewise, people's attitudes towards genetically modified organisms (GMOs) depend on whether a GM plant (or animal) is considered just another variation of the same species, or something very different.

Is the mule-horse story a good metaphor for the GMO controversy? Scientifically it is a horrible metaphor. After all, a GM corn or soybean seed differs from its non-GM counterparts by one or a few genes, whereas the horse and mule are so different they do not even have the same number of chromosomes. It is easy to tell a mule from a horse, but very difficult to tell a conventional from a GM soybean seed.

The metaphor does work for those who fear and oppose GM foods, though. Just as it seems rather "freakish" that a horse can give birth to a mule, GMOs have been labeled "Frankenfoods." There is a GM corn that produces its own insecticide to control rootworms. As a result, animated YouTube videos like "GMO-A-Go-Go" put a Frankenstein-like face on an ear of corn, influencing some to believe that GM foods truly are freaks of nature.

To what extent does the public experience anxiety about GM foods? Polls show that 93 percent of Americans (who responded to a survey) support mandatory labeling of GM foods. This may overestimate true concern, as simply asking people this question suggests there is a problem with GM food. That is, asking the question itself changes the person's beliefs. Think about it: If a telephone survey asks you whether you believe the water you drink to be safe, don't you automatically become suspicious it is not? If there was no problem with the water, why would someone be calling you about it? At the same time, these polls are conducted using valid survey protocols, and it is hard to ignore such a large percentage of citizens.

A different kind of survey was conducted by Jayson Lusk of Oklahoma State University, where 1,004 Americans were surveyed, asking them if they could think of a time they lost trust in the food system. If they answered yes, he then asked them why. This was an open-ended question, so they were not primed by the survey to think of GMOs. About 40 percent said they had indeed lost trust, and of those 413 individuals, GMOs were mentioned 24 times. From this we can say less than 3 percent of survey respondents in the United States are truly concerned about GM food.

The two surveys together tell us that, although very few people have lost trust in the food system because of GMOs, they still want to know if their food contains GM ingredients. So people haven't lost confidence in food because of biotechnology, but they would feel even more confident if GM labels were required.

Before delving into the controversy we should define exactly what a genetically modified organism is. In this book, a GMO will refer almost exclusively to transgenic crops, where the genes from a nonplant organism (usually bacteria) are deliberately inserted into a plant (using recombinant DNA or gene-splicing) in hopes that the new plant will exhibit certain desirable traits, like creating its own pesticide or being resistant to a certain herbicide. Not all GMOs are created in this manner. For instance, a cisgenic plant is formed by inserting into the plant's DNA a gene acquired not from a different organism, but from the same or similar species. A GMO can also be formed by removing or silencing a gene within a plant. Most of the controversy concerns transgenic plants, though we still refer to them as GMOs because that is the term used by food activists.

The transference of genes from one organism to another is nothing new. At least 8 percent of the human genome was transplanted from viruses, but we are not GMOs because this transplant was not the direct intervention of human scientists. The alteration of genes by human intervention is not new either. We intentionally alter the DNA of plants constantly through selective breeding. Genetic mutation is a natural process of evolution, and sometimes these mutations can lead to better crops. The rate of natural mutations can be rather slow, so we sometimes increase the rate by zapping plants with irradiation (scientists often wonder why this form of genetic alteration receives little attention from food activists, while GMOs do).

Technologies in genetic modification are different in that humans are choosing the genes they want to insert into another organism and can create these new plant and animal varieties at a faster rate—and with much more precision. Some say these technologies are our best hope for feeding a growing population. Others say the technologies are used recklessly because of corporate influence in regulation. This disagreement is the GMO controversy.

The controversy is important because GM crops have come to dominate the United States and are spreading across the world. As figure 5.1 shows, three major crops are planted almost entirely in genetically modified varieties. For those who believe GMOs to be advantageous this is a remarkable achievement in agriculture. For those who eat foods derived from animals but are fearful of GMOs the graph is alarming, as virtually all livestock consume corn and soybeans.

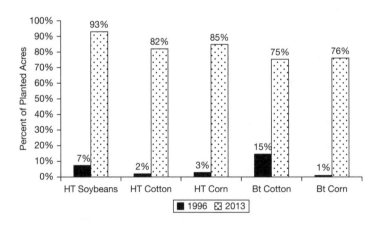

Figure 5.1 Adoption of Genetically Modified Crops in the United States

Note: HT stands for "herbicide tolerant," which means the GM crop is resistant to one or more herbicides. *Bt* signifies a genetically modified crop, engineered to produce its own pesticide.

Source: Economic Research Service, Recent Trends in GE Adoption (webpage), US Department of Agriculture, accessed August 8, 2013, at http://www.ers.usda.gov/data-products/adoption-of-genetically-engineered-crops-in-the-us/recent-trends-in-ge-adoption.aspx#. UgKZBtKQxrw.

How Are GMOs Regulated?

Usually when food processors add something "foreign" to food, like a food additive, unless that additive is deemed to be GRAS (Generally Recognized As Safe) by the Food and Drug Administration (FDA), it is regulated much like pesticides, especially if it is a synthetic additive created in the lab. Before it can be used to, say, color or preserve food, it must undergo a series of rigorous tests to ensure it is safe.

Some believe that taking a gene from one organism and inserting it into another is like adding a foreign substance, and GM foods should undergo similar testing. This type of logic is unworkable in actual regulation, though, because the DNA from all life forms is composed of the same substance. Yet, because genetic modification just seems riskier to many people, a GM seed is not treated the same as a seed created through selective breeding or radiation-induced mutation. A complex system of regulations has been constructed in the United States, where GMOs are vetted by the Department of Agriculture to ensure the crop is safe to grow, the Food and Drug Administration to make sure the food is safe to eat, and the Environmental Protection Agency to verify it will not harm the environment. What follows are the most important features of those regulations in regard to human health, particularly from the FDA.

It is easy to misrepresent how food safety is protected by the FDA. The laws make clear that it is the company's responsibility to ensure food safety, and it is the company's decision whether to consult with the FDA to measure health risks from a new GMO. This system makes it seem that GMOs are not regulated at all, as some critics imply, but this grossly misrepresents how companies actually interact with the FDA. Sometimes there is little difference between a suggestion and a command, especially when the suggestion is made by a powerful government agency, and the FDA has made it clear it wants to be consulted throughout the process of developing any and all GM crops.

FDA believes that it is in the best interests of the regulated industry and the agency for developers to inform FDA…prior to commercial distribution, about foods or feed derived from new plant varieties, including those derived using rDNA techniques.

—Food and Drug Administration, "Consultation Procedures under FDA's 1992 Statement of Policy: Foods Derived from New Plant Varieties," 1997, accessed November 30, 2013, at http://www.fda.gov/Food/GuidanceRegulation/ GuidanceDocumentsRegulatoryInformation/ Biotechnology/ucm096126.htm.

The FDA does not assume that a GM crop variety is safe, nor deny that it may be safe. For this reason the company producing the GM crop will communicate with the FDA throughout the product's development so that the company can answer any questions the FDA poses—and the FDA will have many. What the company wants is confirmation, most often in the form of a letter, from the FDA indicating the agency has no further questions regarding the safety of the new variety. This letter is considered to be "FDA's blessing" by the company, though the FDA would certainly never use that term. Without this letter the company is more vulnerable to lawsuits and can be subject to an expensive product recall by the FDA. Moreover, it is in the company's interest to produce a safe product. Companies don't make money by sickening their customers (at least, not in the long run), so they *want* to work with the FDA to ensure product safety. And because people working for seed companies are just as ethical as everyone else, it follows they also work with the FDA because it's the right thing to do.

For this reason, the GMO "approval" process is best described as a series of consultations that occur throughout a product's development. Its main objective is to determine whether the GM crop is "substantially equivalent" to its non-GM varieties and whether it poses an allergen risk.

"Substantially equivalent" doesn't mean it's safe, only that it's no less safe than non-GM food. There are three criteria by which substantial equivalence is verified. In regards to crops, one criterion is whether the plant looks and behaves like the non-GM plant. Does it mature and flower about the same time, and is it resistant to the same diseases in roughly the same way? These are examples of observational data the FDA might request from the seed company. A second criterion concerns the chemical composition of the final product. For a GM canola variety, for example, the FDA may request information on the seed's triglyceride and fatty acid content. Finally, the third criterion involves information on the nutrients, antinutrients, toxicants, and allergens of the entire plant (even the part not eaten). Because of the variety of crops being genetically modified, there is no one established system of assessing their safety. The FDA reviews each variety on a case-by-case basis, collecting similar data in the beginning but involving different questions as the consultation process proceeds. All of this requires extensive data collection on the part of the seed companies.

If the FDA is concerned about the safety of a GM crop, it may place restrictions on how the crop is used, request animal feeding trials, or oppose the product entirely. Every GM product that has been produced and used in the United States has undergone this consultation process with the FDA (in addition to similar interactions with the USDA and the EPA), so regulation of GMOs is in fact extensive, expensive, and comprehensive. In our opinion, it is also effective.

The concept of substantial equivalence is then the foundation of GMO regulation. Why did the FDA adopt the rule of substantial equivalence? Supporters of the technology will claim that it is because the most prestigious scientific institution—the National Academy of Sciences—supports the notion. The Academy concludes that the *method* by which a plant's DNA is altered is irrelevant, and thus taking a gene from a bacterium and inserting it into a canola seed is not like adding a food additive, but more like selective breeding. So long as the GM canola's DNA is basically the same as its non-GM counterpart, there is no need for additional regulation or testing. So if you ask the most prestigious scientific organization if a GM tobacco seed is basically just another variety of tobacco, they will respond yes.

> No conceptual distinction exists between genetic modification of plants and microorganisms by classical methods or by molecular techniques that modify DNA and transfer genes... the product of genetic modification and selection should be the primary focus for making decisions about the environmental introduction of a plant or microorganism and not the process by which the products were obtained.
> —Committee on Scientific Evaluation of the Introduction of Genetically Modified Microorganisms and Plants into the Environment, National Research Council, National Academy of Sciences, *Field Testing Genetically Modified Organisms: Framework for Decisions,* 1989.

There are some well-qualified dissenting scientists and a motivated group of food activists behind them, pushing back against GM food. They believe a GM crop is not substantially equivalent to traditional crops. Moreover, they believe that the FDA follows the substantial equivalence rule not because of the science, but because the FDA was corrupted by corporate influence. This is not a belief that the authors' share, but there are smart people of high character who do believe this conspiracy theory, and their side of the story deserves to be heard.

In *The World According to Monsanto,* author Marie-Monique Robin describes how the substantial equivalence began with a 1992 policy statement by the FDA under the leadership of a former Monsanto lawyer, who, after working in the FDA, returned to Monsanto as a vice president. Her story suggests that GM regulations were the product of intense lobbying by Monsanto and a revolving-door system where the regulators are former and/or future employees of the company being regulated (note that some argue Monsanto wanted excess regulations to keep out competitors, but that is not Robin's story). It is not hard to imagine a company rewarding lenient regulators with a nice job, and food activists have websites listing powerful government officials and their relation to Monsanto and other corporations. If this sounds like a conspiracy theory (a term not meant as a euphemism), it is.

Consider the 2001 PBS special *Harvest of Fear*, where a representative of Greenpeace claims that the FDA scientists actually advised mandatory labels for GM foods, but that the FDA administrators made a political decision not to. This argument appeared again on a 2013 episode of *Stossel*, where it was clear that the pro-GM side used the credibility of scientific organizations to bolster its case, while the anti-GM side told a political-conspiracy story. With one side telling a scientific story and the other side telling a story of political conspiracy, it's little wonder that the GMO controversy has persevered through decades of debate.

> LUSK: All selection is playing around with genes....In fact,...traditional plant breeding is involving many thousands of genes, and we often don't know what's going to happen. Modern biotechnology is picking one or two genes...so it's actually much more precise than our traditional plant-breeding techniques.
>
> STOSSEL: So that should make it safer. Jeffrey [Smith], what about that?
>
> SMITH: Well, the FDA scientists were absolutely clear in the memo made public from a lawsuit. They said that the process of genetic engineering is different and leads to new and different risks. Like new allerg[e]ns, toxins, and new diseases. They repeatedly urged their superiors to require [more] study, but the person in charge of policy at

the FDA was Michael Taylor—Monsanto's former attorney, later Monsanto's vice president, now is back at the FDA as the US food safety czar.

STOSSEL: So Monsanto has captured the FDA, this thousand-person agency? And [the FDA] is just in the tank with big business?

SMITH: Monsanto has not only captured the FDA, but as I traveled to thirty-six countries, they've done the same to many, many countries....

STOSSEL: ...That makes me skeptical of you [Smith], not them [in that Smith's conspiracy theory is too expansive and intricate to take seriously].

LUSK: You look at every major scientific authority on the subject, whether it's the US National Academy of Sciences, the American Medical Association, the European Commission, the World Health Organization, the Food and Agricultural Organization [of the United Nations]...these are all independent bodies, of independent scientists, and every one of those organizations has confirmed the basic safety of biotech foods.

—John Stossel, host, "War On...," *Stossel*, Fox Business News, June 6, 2013. The guests are Jayson Lusk, agricultural economist at Oklahoma State University, and Jeffrey Smith, Institute for Responsible Technology.

A good conspiracy theory will have a villain who is withholding vital information from the public, and because it is the company's responsibility to perform the research demonstrating the product is safe, there is always the fear that the company is hiding a dangerous secret. Also, much of the information about the effects and performance of GM crops is controlled by the seed corporations, allowing them to release only studies that view GMOs favorably, and suppress studies that do not. Both Marie-Monique Robin's book and the documentary *Ethos* claim that Monsanto deliberately withheld information showing harms from their rBST hormone (given to cattle to increase milk production), and *if* this is true, the anti-GMO crowd wonders, what other data were kept secret?

> At that time Monsanto was saying, "There's no evidence of any adverse effects, we don't use antibiotics," and this clearly showed they had lied through their teeth.
> —Samuel Epstein, interview in *Ethos*, Pete McGrain, director and writer, Media for Action, 2011.

Some may see the above quote as clear evidence of a conspiracy, but this may not be the case. If rBST does what it is designed to do, it will increase milk production, and greater milk production is usually associated with higher rates of mastitis, which requires the use of antibiotics. Both the regulators and the company may have agreed that the antibiotics Epstein refers to are not an adverse effect of rBST at all, and so nothing of concern to the regulators was being concealed.

This hasn't stopped the anti-GM crowd from associating Monsanto with the nefarious activities of cigarette companies decades earlier. A commercial supporting the mandatory labeling of GM foods, aired before Proposition 37 in California, begins by remarking on the scientific support that was once given to the health benefits of smoking, and it is true that cigarette companies deliberately withheld information about the dangers of smoking and manufactured a perception of scientific uncertainty where none really existed.

Scientists have been wrong in the past, and their errors make some discount the scientific support behind GMOs. Scientists once approved the rendering of sheep carcasses for conversion into cattle feed, a practice now linked to the outbreak of mad cow disease. In the case of cigarettes much of the science was deliberately withheld, while in the case of mad cow disease there was no cover-up. Once the link was made between mad cow disease and rendered carcasses, the public was quickly informed and the practice was halted. Though an honest mistake, it shows that scientists can indeed make mistakes. The fear is that scientists are making a similar mistake in regard to genetic modification.

Most of this discussion has concerned the United States, whereas the European Union is far less accepting of GMOs. This difference seems to be attributable to the greater caution exhibited by European consumers, perhaps due to food safety scares in Europe and Britain (especially mad cow disease) that eroded the public's trust in regulators.

> The public response in Europe to GM crops might be very different if the outbreak of BSE, or Mad Cow disease, in the United Kingdom had not occurred in the 1980s. Despite reassurances from the British health officials that consuming British beef was safe, in 1996 the consumption of BSE-tained beef was presumptively linked to a variant form of Creutzfeldt-Jakob disease...the appearance of BSE in cattle in other European countries further eroded the European public's trust that governments were able to assure the safety of food—a trust that had been damaged by a series of food scandals in the 1980s.
> —Karen E. Greif and Jon F. Merz, *Current Controversies in the Biological Sciences: Case Studies of Policy Challenges from New Technologies* (Cambridge, MA: MIT Press, 2007).

Most of the corn, cotton, soybeans, and sugar beets in the United States are GM varieties. Are these really just different varieties of the same crops, or "Frankenfoods?" It depends on the trust one places in scientific organizations like the National Academy of Sciences, and the extent to which one believes that corporations control regulation. As agricultural scientists, the authors have considerable esteem for the Academy and place great trust in the US regulatory agencies, and for these reasons, are supporters of GMOs.

Are GMOs Responsible for the Increasing Food Allergies?

When loving mother Robyn O'Brien gave a talk at TEDx-Austin she told a story of how one of her children developed a severe allergic reaction to a normal breakfast of waffles, yogurt, and eggs. Told these were some of the top foods known to cause food allergies, she thought back to her childhood, when food allergies did not seem to exist. After doing some research, she found that cases of food allergies have exploded in recent decades, with the rate of hospitalizations due to food allergens rising by 265 percent between 1997 and 2002. Could GMOs be responsible? That was certainly O'Brien's claim.

Before a company can market a new GM seed, it is required by government regulation (using the consultation process mentioned earlier) to demonstrate it will not increase the rate of food allergies. The process for doing so is scientific, well established, and we believe, effective. All (known) food allergens are proteins, and the vast majority of them are found in peanuts, milk, eggs, soybeans, tree nuts, fish, crustacea, and wheat. The potential for allergens is assessed by comparing each novel protein that is created by the GM plant (and not present in its non-GM counterpart) to about five hundred known allergens. If a potential exists, further tests must be conducted to ensure safety. Otherwise, it is considered no more risky than non-GM food.

To illustrate, consider a GM soybean seed that produces a particular protein needed by livestock (2S albumin), a protein that is lacking in non-GM livestock feed. Scientists identified the gene that produces this same protein in Brazil nuts, and introduced it into a soybean seed's DNA, thus creating a more nutritious feed for animals. However, Brazil nuts are known to cause allergic reactions in some people, so the company worked with university scientists to use skin pricks to see if people were allergic to the new GM seed. Some people were indeed allergic, and so development of this particular soybean variety was halted and all plant material and seeds were destroyed.

Another example is the GM corn variety called StarLink, which did not pass the allergen test, meaning regulators feared it might cause allergic reactions if directly consumed by humans. StarLink was therefore approved to be used only as animal feed. It did not help the GM cause when StarLink was soon detected in corn eaten by humans, and though its prevalence was still too low to cause problems and no harms have been documented, it showed that restricting how a GM crop is produced is easy to express in words but harder to execute in practice.

The evidence so far suggests that current regulations are providing adequate protection against allergic reactions from GM food. This doesn't mean that reactions won't occur in the future, but there is no reason to expect greater allergic reactions from GM food than non-GM food. Revealingly, the UCLA Food and Drug Allergy Care Center does not even mention GMOs as a source of food allergies. Instead, it attributes the rise of allergies to other factors: the general increase in hygiene over time (the more sterile our environment, the more sensitive our immune system), a delayed introduction of certain foods to children, the increase in processed foods (was O'Brien making waffles from scratch like people used to, or was she using Eggo frozen waffles?), and improved information gathering and reporting. When the *New York Times* printed a commentary among six experts on food allergies, none of them suggested GMOs were to blame. A CNN article titled "Why Are Food Allergies on the Rise?" did not mention them either.

For the present there is little evidence that GMOs are responsible for the rise in food allergies in children. Perhaps O'Brien is just ahead of her time and scientists may prove her correct, but for now her speculations are without evidence. Indeed, if food allergies really are a problem, biotechnology could be used to create new plant varieties that reduce food allergies.

Is Genetically Modified Food Safe to Eat?

When the National Academy of Sciences published its 2004 report on the safety of GM food, it explained that any form of genetic manipulation can have unintended health impacts, but that no health harms attributed to genetic engineering have been documented, nor are they expected to arise. Most other health and scientific organizations agree, including

- American Association for the Advancement of Science
- American Medical Association
- Food Standards Australia and New Zealand
- French Academy of Science
- Royal Society of Medicine
- European Commission
- Union of German Academies of Sciences and Humanities
- Seven other academies of science (Brazil, China, India, Mexico, Third World Academy of Sciences, Royal Society (United Kingdom), and the National Academy of Sciences of the United States)
- World Health Organization

As mentioned previously, when one consults prestigious scientific organizations they generally testify that GM foods are safe, but there are some individual scientists who dissent. Earth Open Source published a book in 2012 arguing that GM food is unsafe, and listed the many animal-feeding trials as evidence. It contains alarming sentences: "Mice fed GM soy showed disturbed liver pancreas and testes function....Old and young mice fed GM *Bt* maize showed a marked disturbance in immune system cells in biochemical activity....Female sheep fed *Bt* GM maize over three generations showed disturbances in the functioning of the digestive system." These are real effects observed in real studies, so why did the National Academy of Sciences say GMOs are safe?

In any study investigating the impact of a certain food source on animals, animal health will vary across groups for reasons other than the feed. Even the most scientific trials contain an element of randomness. For example, for any two groups of almost identical mice, one group will be healthier than the other for no obvious reason. We would not expect all mice to die at the exact same moment. Likewise, even the most tightly controlled experiments require the use of statistics to say what health harms were caused by a particular food and what harms were caused by randomness. Researchers are human, and these judgments are probably impacted by their general beliefs about GM food, but different conclusions also involve simple differences in judgment, absent of bias. Sometimes what appears to be an ideological bias against GMOs could actually be the result of something more esoteric, more mundane, and perfectly understandable.

Likewise, research interpreted to imply that scientists favor GMOs because of corruption also has an alternative explanation. After studying ninety-four articles on the health impacts of GM products, a group of researchers found that scientists who possessed a professional relationship with a GMO company were more likely to conduct research favorable to GMOs (though the source of funding did not seem to matter). Does this prove that industry connections influence research, or could it be that research influences industry connections? A researcher who believes the data clearly show GM foods to be safe is more likely to develop ties with corporations. That belief in the safety of GM foods will then impact the judgment calls made in future research. Corruption may have nothing to do with it. One does suspect that at least a few researchers are swayed by corporate influence (they are humans, after all), but critics of GMOs greatly overestimate their numbers.

The lesson the authors learned writing this chapter is that the GMO debate has become so acrid that it is difficult to have an honest discussion. If a researcher makes positive remarks about biotechnology, she is accused of being a corporate shill, and if one questions the safety of GM crops, she is ridiculed by other scientists. Oddly, treating both sides of the debate with respect only angers both sides. Nevertheless, it is impossible to truly understand the controversy without taking both sides seriously, and that is exactly what this chapter has attempted to accomplish.

Will GMOs Lead to Excess Market Power for a Few Corporations?

It can be difficult at times to see what GM opponents dislike more: genetic modification itself or the large corporations that perform it. When people protest biotechnology, sometimes, they are really protesting the market power of corporations. Creators of new GM crop varieties are given a patent, which is a temporary monopoly on that crop. Patents are no modern creation, but an ancient and reliable system for rewarding creativity. In ancient Greek colonies, cooks were allowed a patent for one year on all new food inventions; seed companies can today acquire a patent for GM seed that lasts twenty years.

> No, my problem with biotechnology is that the science has been hijacked by corporate interests, and that the subsequent wholesale rush to patent plant genes as the intellectual property of a handful of multinational corporations is placing the control of global food production directly into their hands.
> —Andrew Gunther, "GE Crop Thriller Leaves Bond and Bourne for Dust," *Huffington Post*, May 15, 2013.

> The claims made for GM agriculture are a transparent fraud. The real purpose of GM foods is to give giant corporations legally-enforceable monopoly powers over the entire global food chain.
> —Colin Tudge, "The Real Point of GM Food Is Corporate Control of Farming," *Ecologist*, November 1, 2013

This does not mean that big corporations have a monopoly over all crop seeds, though. There are plenty of non-GM varieties of corn and soybeans, but most farmers simply do not want them—they voluntarily purchase the GM seeds. In this sense the seed corporations have earned their market share through the creation of a superior product—just as Google has earned its large market share in search engines through its superior search algorithm. The majority of corn, cotton, soybeans, and sugar beets in the United States are GM varieties for the simple reason that farmers prefer them. Biotech crops are spreading across the world, and because not many firms can sell GM seeds, the four-firm concentration ratio (the percentage of the market dominated by the four largest firms) for the seed and biotechnology sector rose from less than 25 percent in 1994 to over 50 percent in 2009.

It is sometimes said that economists believe a market is no longer competitive when four companies control 40 percent or more of a market, but that's only the case when firms are making an identical product (and even then not all economists agree on the 40 percent rule). Just as Google is a very different search engine than Yahoo's engine, GM seed is not the same product as non-GM seed. These days, seed companies compete less on price than on innovation. Given the constant stream of new seed varieties, one could make the case that the seed market is actually quite competitive.

This does not mean there are no concerns about market power. Five large seed corporations are Monsanto, Syngenta, Dupont, Dow, and Bayer, but these firms do not act independently of one another. Monsanto owns most of the patents for GM seed traits, but other companies have seeds with advantageous traits also. Often, to produce the highest quality seeds possible, Monsanto and one of the other firms will strike an agreement and combine their genetics. For example, Monsanto may own a patent for a GM trait in soybeans, while Dow may have a variety of soybeans particularly suited to the southeastern United States. By combining these traits they can sell a GM seed that grows well in South Carolina, increasing the value of the traits owned by both companies. This is referred to as a cross-licensing agreements. This cooperation requires the companies to work closely with one another, making it easier for them to collude and behave like a single seed monopoly. A Monsanto executive is said to have remarked in 1996 that the entire food chain is becoming consolidated, not just the seed industry. However, if the purpose of cross-licensing agreements really is to create better crop varieties, they may be good for the consumers as well as the companies.

Are We Losing Crop Diversity?

Some feel this corporate hegemony in crop seeds threatens our food supply by reducing the diversity of plant varieties. The Irish Potato Famine (1845–1850) was caused, in part, by the planting of the same Lumper variety of potatoes throughout Ireland. So little genetic diversity existed that virtually all the potatoes were destroyed by one pathogen. To breed potatoes resistant to this pathogen, it was necessary to return to the land of potatoes' origin (South America), where thousands of different varieties were grown. In 1970 a leaf blight struck much of the US corn crop, reducing corn yields by more than 20 percent. The same variety of female corn plants had been used to produce the hybrid seeds that most farmers planted, and the varieties resulting from this cross were particularly susceptible to the blight. Crop breeders quickly learned they needed greater genetic diversity if their seed varieties were to remain popular. The vast majority of bananas come from one variety currently under assault from the Tropical Race Four fungus. The banana case is a particularly interesting example because bananas can only reproduce asexually, creating an extreme uniformity of plant genetics. The lesson is clear: when crops lack genetic diversity a greater portion of the food supply is vulnerable to damage.

> My problem has been less about health and safety of the [GM] technology than it has been about the political economy of GM and what it has done to American

agriculture, to competition in the seed business, and to the size and sustainability of our commodity crop monocultures.

—Michael Pollan, "Pointed Talk: Michael Pollan and Amy Harmon Dissect a GM Controversy," Grist.com, August 28, 2013, accessed August 30, 2013 at http://grist.org/food/pointed-talk-michael-pollan-and-amy-harmon-dissect-a-gm-controversy/.

If a perilous crop disease threatens the food supply the types and varieties of crops planted may need to be altered. It might be prudent to cross (i.e., breed) the most popular varieties with less popular varieties to acquire more genetic diversity. But if these less popular varieties are no longer around because farmers have not purchased them in years, the only way of quickly acquiring diversity is genetic modification, radiation-induced mutation, or the like. Thus, some apprehensive groups have been storing all the different varieties of seeds they can acquire in frozen vaults (in Norway, for instance), an insurance policy that might save millions of lives.

Are we losing genetic variety in our crops? A recent *National Geographic* article documents a decline in variety for sixty-six crops between 1903 and 1983, such that we lost 93 percent of our varieties during that time period. Yet a different study of seed catalogs over that same time period found the number of varieties to have risen for some crops and fallen for others, but overall found no difference—implying we are not losing crop diversity. It is currently unclear which study better represents reality.

Suppose that in the future, genetic diversity in crops is lost, and suppose the change is largely due to successful GM crops. Does that alone put the food supply in danger? Not necessarily. If, say, a horrible bout of rust (a fungal disease) wipes out much of the wheat, one response could be to cross the current wheat crop with more antiquated varieties, hoping some of the crosses would be resistant to the rust. Or scientists at seed corporations could devise a new genetic modification that is also resistant to the rust, which is how the banana industry is responding to the Tropical Race Four disease (but then, they have little choice). Remember the leaf blight that hit the United States in 1970s, mentioned earlier? The problem was solved in only one year, after seed companies—aided by university research—quickly integrated greater genetic diversity into their breeding programs.

It could be argued that to protect the food supply from devastating disease, the most advanced genetic science should be employed, and that would be genetic modification performed by large corporations. One could imagine a world where GM crops are banned and there is greater crop diversity, but a rust is able to infect most of those crops anyway. Seed corporations then claim they can develop a GM wheat seed resistant to this rust in three years. Could it be that genetic modification is the savior? It is a possibility.

Do not be deceived about the true genetic diversity that occurs even for the same type of GM seed. There is no one single Roundup Ready soybean seed, but different varieties with Roundup-resistant genes. Once Monsanto developed a Roundup-resistant soybean, it did not sell that exact same variety to both Minnesota and Texas farmers. Instead, it crossed that GM variety with other soybean varieties best suited to each region, especially the length of a region's growing period. What emerged then were various varieties of Roundup Ready seeds suited for specific settings. Similar stories can be told for all superior varieties, both GM and non-GM alike. When the West Africa Rice Development Associate sought better rice varieties, it took Asian rice for its reputation for high yields and crossed it with African varieties, which are known for their weed-control and drought-resistant traits. This diversity is the norm in plant breeding, and should not be forgotten when discussing seed diversity.

One final comment. Genetic diversity of crops cannot be measured by the market share conquered by the four largest seed corporations. In nineteenth-century Ireland that share would have been almost zero, as most farmers acquired their seed potatoes not from seed companies but last year's crop from their own fields. Still, farmers managed to plant a similar crop across all of Ireland. If a large seed company had entered the country and tried to acquire market share by selling better varieties, that company's quest for profits might have averted a famine.

Should GM Labeling Be Mandatory?

The United States allows GM food to be sold without a label, though firms can always voluntarily label their food if their consumers value it. Whole Foods made this move recently when it announced that by 2018 its food would be labeled GM or non-GM. What Whole Foods seeks to accomplish by 2018 the European Union established in 1997. All food from GM sources sold in the European Union must be labeled as such (except milk, meat, and eggs fed from GM feed, and a few other categories). Europe has gone beyond labeling, placing tighter restrictions on the planting of GM crops and allowing individual member countries to ban GM products if they wish. A number of EU nations like France and Romania do not allow GMO maize to be planted.

Why do the United States and the European Union view GMOs so differently? One explanation is that European producers sought the labeling law as a trade barrier to benefit European companies. There is little evidence that such indirect trade restrictions have benefited European farmers. In the 1990s much of EU agriculture actually supported biotechnology. Europe treats GMOs differently than the United States because European consumers view it differently. For example, roughly two-thirds of American consumers in the 1990s supported GMOs, while a similar proportion of French consumers opposed it.

A movement has emerged in the United States to require food manufacturers to label all food according to whether it contains GM ingredients. Supporters of labeling argue that consumers have a right to know. There is much appeal to this argument. It is only information, after all, allowing consumers to decide for themselves whether their food contains GM ingredients. Even advocates of biotechnology sometimes support labeling, arguing that attempts to oppose it seem like an effort to hide something.

> By fighting labeling, we're feeding energy to the opponents of GMOs. We're inducing more fear and paranoia of the technology, rather than less. We're persuading those who might otherwise have no opinion on GMOs that there must be something to hide, otherwise, why would we fight so hard to avoid labeling?
> —Mark Lynas, "Why We Need to Label GMOs," speech at the Food Integrity Summit, October 15, 2013, accessed October 18, 2013, at http://www.marklynas .org/2013/ 10/why-we-need-to-label-gmos/.

Labeling opponents pose two arguments. One is that requiring such a label will give the false impression that GM foods are unsafe. It reminds one of when, in the sitcom *Arrested Development*, Gob Bluth suggested his construction company adopt the slogan, "The Bluth/Morento Company: A Columbian cartel that *won't* kidnap and kill you!" stating, "Underline 'won't' because that makes the competition look like maybe they [will kidnap and kill you]." Indeed, the groups pushing for the mandatory labels are also very keen on making GM foods look bad. If a label were passed and it destroyed the market for GM crops, the Food Democracy Now! organization would no doubt celebrate and consider the labeling bill a success.

The other argument against mandatory labeling is that if consumers really wanted the label, then companies would voluntary adopt it, as Whole Foods plans to do by 2018. Let markets, not food activist groups, decide how food is labeled, they say. This sounds reasonable, but one must wonder whether we would currently have nutrition labels telling us the amount of calories, fat, and sugar in foods if it were not for the Nutrition Labeling and Education Act (NLEA) of 1990. Most everyone supports these nutrition labels today, but the food industry was unwilling to provide them until they were mandated by law.

This issue has not divided people among their typical ideological camps of pro- or anti-regulation. Some regular supporters of government regulation—like President Obama's former regulatory czar Cass Sunstein—oppose mandatory GM labels. State Representative Harvell of Maine, a Republican, supports a labeling bill for Maine based on the idea that markets need information to work well, and he quotes the libertarian hero Ludwig Von Mises to defend his position. Others who normally announce no political opinion, like Charles of the United Kingdom, have publicly asserted that GMOs don't just pose health harms but threaten the world's ability to feed itself.

Strangely, both sides of the labeling debate claim to have consumers' best interest in mind. One side observes that polls show strong support for labeling, while the other side retorts that when California actually held a vote on labeling, it failed to pass. This support might fall in real elections because consumers take the issue more seriously, because the people who vote are different from the people who respond to polls, or because people are swayed by the political advertising of biotech companies. One side protests that consumers have a right to know what is in their food, and the other side rebuts that consumers don't want everything about the product on the label, but only the information that matters. If only there was a way to ask consumers what they "really" want, but the only thing clear is that consumers say different things in different contexts. It is almost as if consumers themselves are unsure of what they want.

Will GMOs Help Us Feed the World?

A typical defense of GM crops and livestock first asserts that agricultural productivity gains must continue in order to feed the extra 2 billion humans expected by 2020, which some estimate will require 40 percent more food than produced today, and that current productivity trends are not optimistic in meeting these goals. Then the argument postulates that biotechnology, which includes genetic modification as well as other tools, should be one among many technologies available for reaching this goal.

There is a logic to this argument. Between 1950 and 2000 the world population rose from 2.5 to 6 billion people, yet the only famines that occurred were largely due to political causes, like the central planning failures in China and the dictatorship in North Korea. Those not living under repressive regimes were mostly able to eat, thanks in part to the Green Revolution. This was not a political revolution but one of agricultural science, where new plant breeding and chemical fertilizer techniques allowed food production to increase faster than the world population. Technology saved the day, it seems. Will GMOs be the technology that saves us in the coming decades?

Are GM crops more productive? There is no reason to believe farmers using GM crops should have higher yields, as the farmers who first adopt GM crops are different types of farmers than those who do not, and their yields are influenced both by the productivity of GM grains as well as their land and managerial skills. These nuances are observed in the United States where insect-resistant corn has increased yields whereas herbicide-resistant soybeans have reduced yields (though only slightly).

Only controlled experiments can isolate the effect of genetic modification on yield, and some of these studies find that GMOs increase yields while others identify a decline in yield. Yield is important but it is not everything. A lower-yielding GM variety may still be preferred if it reduces pesticide costs, is more resistant to drought, or helps control pests for a different crop planted subsequently (e.g., planting GM canola this year to help control weeds in next year's non-GM wheat).

Genetic modification is just one tool for producing a better crop variety, and if one GM crop isn't successful that doesn't mean other GM crops won't be. The Flavr Savr tomato was the first GM food, and it failed, but it was later followed with soybean and corn varieties that would dominate the market. Any time technology providers and farmers have more options in how to produce food they will produce more efficiently—otherwise, they will be driven out of business by those more efficient. If one has faith in the regulatory system that no GM food will be approved unless it is safe and environmentally-friendly, then it is hard to imagine how GMOs cannot help us feed the world.

Anna Lappé with the Food MythBusters organization argues otherwise, saying that we have been tricked into believing that technologies sold by corporations are necessary to feed the world in 2020. Her argument is that these technologies initially seem advantageous but farmers develop a "quick addiction" to GM seeds, fertilizers, and pesticides. Eventually the soil becomes depleted and pesticides become ineffective, thus threatening our ability to produce food, she predicts. This prediction is based on her belief that corporate influence has "tilted the playing field" to favor large corporations and thus the technologies they sell, like GM seed.

Once again, whether one believes GM foods can help feed the world in 2020 depends on one's view of corporations. If, like Lappé, one believes that corporations can reduce farmers' options by controlling regulation, the input market, and the output market, then they hinder our ability to feed the world. Conversely, if one believes GMOs are regulated effectively and corporations can only succeed in the market if farmers consistently prefer their product, then GMOs increase farmers' options and might play a major role in feeding future generations.

Do GMOs Reduce Pesticide Use?

If GMOs can reduce pesticide use then there might be health and environmental benefits from the technology. Both sides of the GMO divide seem to agree that crops that create their own insecticide (e.g., *Bt* corn) have lowered insecticide use. It would have been shocking if they did not. That said, the use of any insecticide, whether it is sprayed by the farmer or created by a GM plant, leads to pest resistance, making the pesticide less effective. Lately, the corn rootworm has become resistant to the insecticide produced by *Bt* corn, such that farmers are now having to spray insecticides again. If this continues, then the reductions in insecticide use may be reversed.

But what about the pesticides in the plant itself? If the EPA considers the *Bt* corn to be an insecticide (it does), it should certainly take into account the *Bt* insecticide produced by the plant in calculating total pesticide use. At the time of this writing it is unclear how total insecticide use trends would change if the plant's own insecticide is accounted for. Although some sources claim the *Bt* toxin is thousands of times more concentrated in GM plants than pesticides containing *Bt* (like Foray 48B) this is something EPA would take into account when determining whether *Bt* crops are safe, so it is doubtful that *Bt* products lead to increased exposure to pesticides.

There is a disagreement regarding GMO's effect on herbicides. One line of research shows total herbicide use decreasing because of GM crops, while another measures an increase. So, do GM crops like Roundup Ready cotton lower total herbicide use? It is unclear. One thing known is that the toxicity of Roundup is very low relative to the herbicides it replaced, so even if herbicide use has risen, the total amount of harmful substances applied to cropland—expressed in units called the Environmental Impact Quotient (EIQ)—has probably fallen, benefiting the environment and consumer safety.

Are GMOs Good for the Environment?

How can GMOs benefit the environment in ways other than pesticide use? If GM crops are more productive, then farmers can produce the same amount of food with fewer inputs, which means smaller amounts of pollution, better resource conservation, and more land available for wildlife preservation (higher yields per acre means less land per calorie produced). As a headline from an article in *The Economist* reads, "Frankenfoods reduce global warming." GMOs additionally make no-till agriculture more feasible, thereby reducing soil erosion and sequestering carbon from the atmosphere. From the previous section we saw that—although GM technologies often increase yield—even if GM crops have lower yields they would be accepted by farmers if they are more efficient, and more efficient means fewer inputs, fewer resources, and less land used to produce food.

GMOs in agriculture have focused on lowering the cost of agricultural production, rather than social problems like reducing water pollution from fertilizers. The reason is clear. Corporations operate on money borrowed from investors, so they must make sure all their activities are devoted to paying back those investors. If investors wanted that money to help clean America's lakes and rivers, they would donate it to a nonprofit organization instead of buying corporate stocks. But they didn't, and so we should not expect Monsanto to try to save the world, but only generate products that other people will pay for.

The presence of avian influenza and the world's huge populations of chickens—especially those raised outdoors, where they come into contact with feces of other wild birds—presents a serious health threat. Scientists have developed a GM chicken that is immune to and does not spread the deadly virus to other chickens. Because such a virus can spread quickly around the world, one can only imagine how many lives such a chicken could save.

There are endless other ways to achieve public goods through genetic engineering, and we haven't even mentioned the GM plants and animals used to produce human organs and pharmaceuticals. The reason GMOs tend to be associated largely with corporations is that funding for biotechnology in the public sector is declining, and this is partly due to the controversial nature of GMOs. If another decade passes and no harm from GM foods is proven, opposition by food activists will either wane or become irrelevant to the rest of society. Then perhaps biotechnology can accomplish social goals like a cleaner environment and disease prevention. In the meantime, though, the controversy becomes more acrimonious every day.

6

THE FARM SUBSIDY CONTROVERSY

What Is the Farm Subsidy Controversy?

The last twenty years have witnessed a stunning rise in food activism, where authors like Michael Pollan, activists like Alice Waters, politicians like former mayor of New York Michael Bloomberg, and over twenty food documentaries at Amazon Instant Video are trying to convince average people that they eat unhealthy food. Agricultural economist Jayson Lusk playfully refers to these individuals as the "food police." His book titled *Food Police* is dedicated to "those who wish to eat without a backseat driver."

Of course, the activists do not want to seem like backseat drivers. They argue that big business and big government are the real backseat drivers pushing upon the public an unhealthy diet.

What are these "unhealthy" foods abhorred by the food police (we also use the term with humor)? Corn is the big one. Corn itself isn't considered unhealthy. It is simultaneously a fruit, a grain, and a vegetable—hardly unhealthy foods. However, most corn is eaten not by humans but by livestock. Corn seems to be in virtually every food product, whether it be corn sugar in soda, maltodextrin in granola bars, or citric acid in canned fruit. Trying to find a processed food without corn is difficult. Because most health experts suggest eating less meat and processed foods, and corn is often an input into less healthy foods, it is thus a target for the food police.

The documentary *Food, Inc.* takes the viewer to a grocery store where an endless variety of products are on display, and then remarks that this variety is an illusion—that everything is traced back to corn. The viewer is then taken inside a combine as corn is harvested, hearing the farmer explain that he produces corn rather than any other crop mostly because of farm subsidies (and the corporations that lobby to receive those farm subsidies).

> In the United States today, 30 percent of the land base is being planted to corn. That is largely driven by government policy that, in effect, allows us to produce corn below the cost of production. The truth of the matter is we're paid to overproduce. And it was caused by these large multinational interests. The reason our government's promoting corn is the Cargills, the ADMs, Tyson, Smithfield—they have an interest in purchasing corn below cost of production. They use that…extensive amount of money they have to lobby Congress to give us the kind of farm bills we now have.
>
> —Troy Roush, vice president of the American Corn Growers Association, interviewed in documentary *Food, Inc.* "Purchasing corn below the cost of production" requires that government subsidies to growers make it possible to sell corn at a lower price than it costs to produce

US food activists tout beans as one of the most delicious and nutritious foods, but not soybeans, because, like corn, most of the nation's soybean crop is fed to livestock and used in processed foods. Even those who realize that meat consumption in moderation does not endanger health admit that Americans should consume more vegetables. Moreover, if high consumption of meat, eggs, dairy, and processed food is really due to farm subsidies, there is indeed a cause for concern. The first issue is the extent to which farm subsidies are the reason so much land is planted in soybeans and corn, instead of plants like arugula and squash.

Then there are the farm subsidies themselves. Why do we have them? What do they accomplish? Are they intended to benefit large corporations, small farmers, or consumers? These are the controversies we confront, though the discussion concerns primarily subsidies in the United States.

Are Farm Subsidies Responsible for our Enormous Corn and Soybean Production?

More cropland is used to produce corn and soybeans than any other crop, and they are used mostly for livestock feed and processed foods. That is, so much corn and soybeans are grown because consumers eat so much meat, dairy, eggs, and processed foods. Because food activists generally want to reduce consumption of meat, dairy, eggs, and processed foods, they believe human health would improve if these corn and soybean acres were planted in something else (radishes or broccoli, perhaps?). Food activists especially focus on corn, though.

Do subsidies cause excess corn and soybean production, and is that why we eat so much meat, dairy, and eggs? Is that why corn finds its way into almost every processed food? Farm policies are incredibly complex in the United States (and the European Union). They don't just subsidize farmers for each unit of a commodity they produce. Sometimes they give this subsidy while also limiting the amount of the commodity they can produce. At other times farmers are given a lump sum of money regardless of what or how much they plant, which shouldn't (directly) affect planting decisions at all. There are even policies that restrict imports of goods like sugar, which serve to increase prices and decrease consumption. The simple existence of programs we call "subsidies" doesn't mean excess production takes place. There are many other reasons why corn and soybeans have become the dominant agricultural crops in the United States. People really like meat, especially from corn-finished cattle, and a corn/soybean ration is a great food for chickens and pigs. Our love for meat, not subsidies, could be the answer.

Or technological innovations in corn and soybeans might have been more pronounced for these two crops than other foods, making it cheaper to raise them. Fortunately, a simple thought experiment can clarify the role of subsidies versus technology. Imagine a world where there are no productivity gains in corn but large subsidies are given. As corn production expands in response to subsidies, the new acres will be less productive than acres already in production. After all, wouldn't farms utilize the best land first? If this is the case, then the average productivity of agriculture would be falling (average yields would be falling) as corn acreage expands.

Then imagine another world where there are no subsidies but there are substantial productivity gains. This is a case where agriculture becomes more productive as corn acreage expands. In reality, subsidies and technological innovation have occurred simultaneously, but which one has the greater effect? This can be answered by studying whether productivity has fallen or risen in the last seventy years.

Figure 6.1 depicts corn production and yields from the 1920s to the present. Both corn production and corn productivity have been rising steadily, in tandem, over time. This rise in production isn't from farming more acres, but getting more out of each acre. Similar trends have taken place for almost all of agriculture, but productivity gains have been especially remarkable for corn.

Agricultural economists have long studied the relationship between farm policy and farm production, and they generally find the subsidies (we define subsidies here as any program that delivers monetary benefits to farmers, even if indirectly through import restrictions) have very little effect. Looking back over the last century, the late Bruce Gardner, one of the

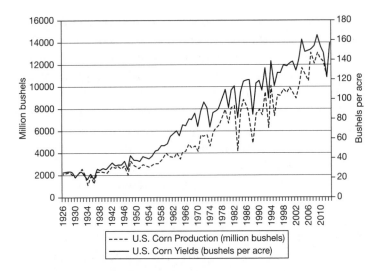

Figure 6.1 US Corn Production and Yields

Source: Economic Research Service, "Corn: Background," July 17, 2013, accessed July 23, 2013, at http://www.ers.usda.gov/topics/crops/corn/background.aspx#.Ue6kJ9KQxrw.

most respected agricultural economists, concluded that farm subsidies really play a very small role in shaping the behavior of farmers.

> A wide range of models is consistent in yielding fairly small output effects…from U.S. commodity programs.…Comparisons…provide no support for the idea that commodity programs have made a difference [in the amount of agricultural production].
>
> —Bruce L. Gardner, *American Agriculture in the Twentieth Century: How It Flourished and What It Cost* (Cambridge, MA: Harvard University Press, 2002), 347–48.

Other economists analyzing more recent policies continue to find that the subsidies have little impact on how much farmers produce. The Federal Crop Insurance program has some effects, but they are very modest. Those who analyze the effect of decoupled farm subsidies (where farmers are given a lump sum of money no matter what or how much they plant) does influence farmers, but once again the effects are small. Research by Bruce Babcock shows that if all subsidy payments related to corn and soybeans were eliminated, prices would rise by no more than 7 percent.

We can't finish this section without discussing one of the most famous secretaries of agriculture: Earl Butz. He never liked the idea of giving farmers a lump-sum payment regardless of how much they produced. Agriculture exists to feed people, he believed, and it should produce food efficiently so that consumers pay low prices. Serving under Nixon and Ford, he made a famous speech where he said farmers should plant as much as they could, and that if they overproduced he would find an export market to sell it. That he did, negotiating a big grain sale to the Soviet Union in 1972, helping the USSR cope with a major drought.

He also urged farmers to "get big or get out," which might seem unkind to small farmers, but as an economist he understood economies of scale, and was ahead of his peers in recognizing the efficiency advantages of larger farms. Studies would eventually show Butz to be correct in saying larger farms would be more efficient. He thus believed that some farms would grow larger and outcompete smaller ones, and tried to help farmers plan for their future by understanding this fact. Because he knew these bigger farms would make food cheaper, he believed them to be in society's interest. His opponents might label him as pro-corporation, but his supporters could say he was pro-consumer.

Because Dr. Butz stressed the importance of increasing the food supply, he is blamed for our heavy reliance on corn, but increasing corn consumption and production are not the same things. In fact, the corn exports he negotiated increased corn prices in the United States, which would serve to decrease—not increase—corn consumption in the United States. Also, observing figure 6.1, showing corn production from the 1920s until today, there are definite ups and downs, and that production does trend upwards during the 1970s, but the general trend is no different between 1960 and 1972 than it was from the 1970s forwards. If Butz did cause corn production and consumption to rise from his sheer personality alone, it is hard to detect in data.

Again and again the conclusion is that subsidies administered in the twentieth century only increased corn and soybean production modestly. The reason is that the subsidies were more complex than simply a fixed amount of money given for each bushel of corn or soybeans harvested. Sometimes farmers could only receive them if they agreed to limit the amount of crop they produced. Other times the farmer received a higher price because imports into the United States were restricted. In a sense, saying the "subsidies" didn't do much is a bit deceiving because there was rarely any program that could be described simply as a subsidy.

Until ethanol, that is. Ethanol is a biofuel made from corn. In the past, ethanol producers received a 45-cent subsidy for every gallon of ethanol produced and were protected from international competition by an ethanol tariff. Although these subsidies and tariffs have been removed, the US government still subsidizes ethanol indirectly by requiring a certain percentage of gasoline to be blended with "renewable" biofuels like ethanol, and this percentage might rise in future years. This is one subsidy that may have increased corn production.

Subsidies for ethanol have existed for over thirty years, but they increased considerably in 2005, after which total corn production rose and corn prices ascended to a new high. Although the subsidies may have been given directly to the ethanol producers, corn producers received a large share as the price of corn was bid up. Though relatively little ethanol was produced in 2005, by 2011 more US corn was used to produce ethanol than was fed to livestock (though some of the ethanol by-products are then fed to livestock). Ethanol subsidies appear to have influenced what farmers planted, causing them to grow more corn, less soybeans, and raise less livestock. This is not an outcome the food activists should oppose, though, because it made both corn and foods derived from livestock more expensive.

Do Farm Subsidies Cause Obesity?

It has become a common notion that obesity could be curbed by a greater consumption of fruits and vegetables. Even if farm subsidies cause the price of grains, meats, and processed food to fall only slightly, perhaps eliminating them will reduce consumption of these "bad" foods and increase fruit and vegetable production. Hypotheses like this have caused some to blame farm policy for today's rise in obesity. Is it true?

Research suggests that removing farm subsidies of grains like corn and soybeans would reduce caloric consumption, but the average adult weight would decline by only 0.35 pounds per person per year. Removing indirect subsidies like the import quotas on sugar would actually increase obesity slightly by making sugar less expensive, and if we remove all direct and indirect farm subsidies the average adult weight would rise, but by less than 1 pound.

Studies attempting to quantify the precise change in weight due to an alteration in public policy requires a vast simplification of the real world and represents more of a thought experiment than an accurate projection, so the actual effects of removing farm subsidies might differ from projections, but there is no compelling evidence that doing so would have much influence on obesity. Australia has eliminated its farm subsidy programs but displays the same pattern of obesity as the United States, and there isn't any reason to believe the American experience would be different.

Why Do We Have Farm Subsidies?

At this point the reader may stop to ask why we even have these farm subsidies in the first place? Critics of US farm policies usually lament that they were initially designed to help small, struggling farmers during the Great Depression, but since then they have evolved to favor rich people and multinational corporations. There is some truth to this, but farm policies have always been a political strategy used by politicians, and those occurring after or before the Great Depression are no exception. Politics predates the Great Depression by millennia.

President Franklin Delano Roosevelt (FDR) was elected to the presidency in 1933 on promises that he would actively fight the nation's prolonged recession. His administration believed that markets had failed to coordinate agricultural activities effectively, and that government could do better, so part of his New Deal legislation was to establish a formal and somewhat Byzantine set of policies that would allow government to control prices, make loans to farmers on favorable terms, control production, store commodities, and provide insurance. As these policies were being implemented, many politicians tried to decipher FDR's formula for how much government support each state would receive. What they found was that most money was devoted to swing states, not because they needed more relief but because FDR wanted to ensure their support in the next election. This doesn't mean that there were no altruistic intentions behind the New Deal legislation, only that altruism was partnered with politics, and politics has a way of taking over.

If you ask agricultural economists today why we have farm subsidies, few will say they are intended to help struggling farmers. Most, especially those who study farm policy, will remark they exist for political reasons. There is little doubt that we restrict sugar imports because it makes the Fanjul brothers rich, and the brothers give handsomely to both political parties (they have so much influence that Bill Clinton interrupted breaking up with Monica Lewinsky to take their phone call!). Julian Alston and Daniel Sumner are among the most respected agricultural economists today, and they say plainly that the real purpose of agricultural policies is to redistribute wealth from the taxpayers at large to a targeted group of individuals. The reason is simple. Politicians who take very small amounts of money from many taxpayers and give it all to a few people will not anger the taxpayers but will cause the recipients to be very gracious—and that gratitude will be expressed in campaign contributions.

> [Farm subsidies] were never designed to be subsidies to help poor people.... The only decent reason to have these subsidy programs is because we've always had them. There's no other reason you can think of.
> —Daniel A. Sumner, "Agricultural Subsidies: Corporate Welfare for Farmers," ReasonTV.com, interview by Nick Gillespie, January 27, 2009, accessed June 3, 2013, at http://www.youtube.com/ watch?v=TeAYuLB8VTg.

Ethanol subsidies were ostensibly created to benefit the environment. Environmentalists may have backed them at first but now seem opposed to ethanol. *Rolling Stone* magazine published an article titled "The Ethanol Scam," claiming ethanol harms the environment, and though there is disagreement about whether this is the case, environmental groups now show little support for ethanol. Everyone seems to dislike ethanol now, except corn and ethanol producers.

Like traditional farm subsidies, the real origin-story of ethanol subsidies is politics. Al Gore has confessed that any environmental benefits are trivial, and when he explained his past support for ethanol he said the following:

> One of the reasons I made that mistake is that I paid particular attention to the farmers in my home state of Tennessee and I had a certain fondness for the farmers in the state of Iowa because I was about to run for President.
> —"Al Gore's Ethanol Epiphany," *Wall Street Journal*, November 27–28, 2010, A16.

Farm subsidies, then, began partly for political reasons and continue to exist mostly for political reasons. Let us leave aside the intentions behind the farm bill and observe its outcomes. Farm subsidies flow mostly to a select few. From 1995 to 2012, the top 10 percent of subsidy recipients received 75 percent of all USDA subsidies. During this period, the families of twenty-three members of Congress received farm subsidies, and one representative from Tennessee received over 3 million dollars. Almost all of the subsidies go to corn, rice, cotton, wheat, and soybean farmers, leaving little for those who produce fruits and vegetables. While large farmers do receive more total dollars from farm subsidies, relative to the value of their output they receive about five times less than small farmers. As a result small farmers owe more of their total farm income to subsidies, even if they receive less total dollars compared to large farms.

About 45 percent of US cropland is rented, and much of the farm subsidies money ends up in the hands of the landowner, even if the government check is written expressly to the farmer. If the government begins sending farmers more money for every bushel of corn produced, landowners will realize farmers are making more from the land they rent. The landowner is then able to increase the rent, and this is exactly what happens. Landowners are able to exert such pressure on farmers because good cropland is fixed in quantity and cannot be easily expanded, whereas farmers willing to rent land are relatively plentiful. Wealth generally flows to those with the fixed resources, and in this case, they are the landowners. To what extent does the transference of farm subsidies actually take place? It depends on the source you look at. Some sources indicate the landowners receive most of the subsidies through higher rental rates, while others suggest they take only around 25 percent.

Farm subsidies have usually been ignored in the past but lately have become the focus of heated debate. Now that environmental groups are focusing on pollution from agriculture, and citizens are more interested in food, farm subsidies are increasingly criticized. The Environmental Working

Group (EWG) has compiled a database of each individual who receives farm subsidies to increase public awareness, and as the 2013 farm bill was being debated the EWG counted more than 630 editorials arguing we need to rethink how we administer farm subsidies—and whether we should provide them at all. Ethanol subsidies continue to be opposed by everyone except ethanol and corn producers, but the US government continues to require that gasoline contain a certain amount of ethanol.

The United States may be at a point where farm subsidies will undergo a serious change. With the Tea Party putting greater pressure on Republicans to curb government programs, environmental groups proving to be formidable challengers to the subsidies, and farmers being less concerned about receiving subsidies in the presence of spectacularly high grain prices, farm subsidies might be a fading institution. This is said in the summer of 2013. Yet everything may change within a year, and it appears that farm subsidies might evolve into a subsidized crop insurance program. Whether these subsidies are larger or smaller than their predecessors remains to be seen.

The United States is not alone in questioning farm subsidies. Rice farmers were promised large subsidies in Thailand to win their votes, but when this promise was fulfilled the government discovered the program was too expensive and was unsure what to do with the surplus rice it accumulated. Indian politicians are trying to reduce the fertilizer subsidies the government doles out, both because of high costs and because farmers are applying so much they are harming the soil. Food and agricultural subsidies comprise 4 percent of Egypt's budget, making the price of food so low that people feed cheap bread to animals.

Developing countries have an even harder time reducing subsidies than the United States or European Union because low food prices are seen as necessary to political stability. US politicians may fear losing campaign contributions if they remove subsidies, whereas their Third World counterparts may fear losing their heads.

7

THE LOCAL FOOD CONTROVERSY

What Is the Local Food Controversy?

As Romania exited the Soviet bloc in 1989 and began integrating with western Europe, it returned the land comprising its collective farms to its owners prior to collectivization (or their heirs). As the country transitioned towards a market economy, the region of Transylvania did the unexpected regarding milk. Instead of relying on inexpensive milk produced by modern and distant farms, they developed a market for local milk and a special reverence for traditional farming methods. Local milk was valued more not simply because it was thought to be of higher quality, but because it represented a traditional culture Transylvania did not want to see wither in the wake of globalization. This culture has a long tradition of rearing livestock, and some people suspect (due to their high rates of lactose tolerance in adulthood) the Transylvanians were among the first to consume sheep milk.

Instead of adopting modern machinery to cut and bale hay from large tracts of land quickly, farmers grasped their handheld rakes and made hay as their ancestors did hundreds of years ago, hay they would then transport with a horse and cart to a barn attic. A farm of only eight acres and a few milking cows is typical, and though the system is inefficient by modern standards, it provides about 60 percent of the country's milk.

Farmers milk their own cows and then transport the milk (often in buckets) to a cooperative where it is mixed with other local milk and sold at much higher prices than nonlocal milk. Why? The main reason seems to be a fondness for traditional culture, and it has as much to do with the landscape as it does the milk quality. Transylvanians adore a hayfield teeming in color and plant diversity, and their vocabulary has many more terms to describe landscape than other cultures. Pesticides and chemical fertilizers are disliked partly because they kill the flowers and other plants that grow naturally with grass in the meadows. When Transylvanians buy local milk they acquire far more than the milk itself: they are paying to preserve their past. Modern agriculture with its large and efficient production methods may produce cheaper food, but it changes the community, a change Transylvanians avoid by willingly paying a higher price for local milk. Because they believe in the traditional way of producing milk, they believe this local milk to be "real whole milk," though it is unclear the extent to which they believe the milk to be of higher quality.

> Because it is real whole milk…a piece of the past which their city life has left behind.
> —A Transylvanian's answer as to why cities were paying higher prices for local milk. Adam Nicolson, "Hay. Beautiful," *National Geographic*, July 2013, 124.

This rather romantic picture of local foods is not just held by Transylvanians. For many of the same reasons, some Americans and Europeans refer to themselves as locavores, meaning they prefer to purchase food from small, nearby farmers. For some this means visiting farmers' markets, or Amish food markets where, like the dairy farms in Transylvania, antiquated farming methods are used. Others belong to a network of community-supported agriculture, where one becomes a part owner in the farm. Members pay a subscription fee and receive a share of whatever the farm is harvesting at the time.

The qualities of local foods are rarely questioned, and are usually assumed from the start to be more healthy and nutritious. This seems to be the case at colleges. Our university holds an annual creativity award, which in 2008 went to two individuals proposing a "farm to university" dining program. The proposal was selected because it was assumed to benefit the local economy and environment, not because it was shown to.

> My university has a sustainability coordinator whose main message, as far as I can tell, is to go out and tell people to buy food grown locally...Why? What's bad about tomatoes from Pennsylvania as opposed to Ohio?
> —Richard Vedder, "The Real Reason College Costs So Much," *Wall Street Journal*, August 24–25, 2013, A9.

Before Michael Pollan, there was the author Wendell Berry, who expressed an admiration for the traditional farming styles used by the Amish and urged us to develop closer attachments to local farms, promising a stronger local community would blossom. Food, Berry claims, cannot be separated from the region it is grown, for when you purchase a food item you are indirectly approving of the economic system in which it was created.

There is no controversy about an individual wanting to develop an attachment to local agriculture. The controversy begins when locavores attempt to argue that local foods are superior in all ways to nonlocal foods. They claim that eating local foods helps the local economy prosper. They claim it is better for the environment. They claim local foods are healthier. They make all these claims with very little evidence. We now observe these three controversies and then conclude with our perspective of what local foods really represent.

Are Local Foods Healthier?

There are some advantages and disadvantages of local foods. If you can find fresher, better-tasting fruits and vegetables from a local source (like a farmers' market) than a grocery store, then local foods can be a proxy for higher quality foods. Most foodies will attest to the fact that the best tomatoes are always found at farmers' markets. Because these fruits and vegetables undergo very little processing and are sold locally, there are no big machines and factories, and no massive distribution system is needed. It is anything but "corporate" food, and for some individuals this means a lot.

It would be a mistake, however, to assume that all local foods are healthier. Frozen fruits and vegetables are only slightly less nutritious than their fresh counterparts, while often being cheaper, more convenient, and more available. Canned foods can still be very nutritious, and given their lower cost may provide some households with better access to healthy foods than relying on local, fresh, or frozen produce. While it is common to deem all processed and precooked foods as unhealthy, let us not forget the contribution frozen dinners like Weight Watchers Smart Ones and Lean Cuisine have made to helping people lose weight. Let us acknowledge that the salads at Chick-fil-A are impressive in their taste, variety of greens, and nutrition. Local foods may be on average healthier than nonlocal foods— we don't know, and it depends on the time and place. What is certain is that labeling nonlocal food as "unhealthy" is unfair.

To the extent that local foods are healthier and better tasting, the rise of the local food movement is a step in the direction of better food. In the sections below we critique two claims often made by locavores regarding economic development and the environment, but we urge readers to remember that these critiques are separate from the issue of food quality. Even if the following two sections make readers skeptical about local foods' ability to enhance the local economy and the environment, readers are still justified in buying local, so long as they believe the food is of higher quality.

Do Local Foods Have a Smaller Carbon Footprint?

The answer is: it depends. Local food travels shorter distances between farmers and consumers, and so these smaller number of "food-miles" can result in less carbon emissions, leading some to believe that local food is better for the environment. Fewer food-miles does not mean less fuel consumption, though. A car with a hybrid engine may use less gas to cover 100 miles than the same car with a nonhybrid engine covering only 75 miles, simply because the hybrid engine is more fuel-efficient. Likewise, even if Kroger grocery stores must cover more miles to deliver the same amount of lettuce to stores across the country (compared to a system where each store obtains lettuce from local farms), they can better afford efficient trucks and are less likely to send trucks partially loaded. The Economic Research Service conducted case studies of various food distribution systems, and sometimes local foods resulted in less fuel per pound of food shipped, and sometimes did not. In some cases, local foods required fewer food-miles but resulted in higher fuel consumption (per pound of food).

Imported food could even have fewer food-miles. Though food may travel fewer miles from the farm to the farmers' market, consumers must travel extra miles to patronize the farmers' market in addition to the grocery store, and these extra miles can result in a larger carbon footprint. Moreover, since personal automobiles are relatively inefficient compared to large tractor trailers, the best way to reduce carbon emissions may be to transport food from many distant locations to one grocery store, rather than have each shopper drive to many different local food outlets.

If we truly care about our carbon footprint, we should be concerned with carbon emissions observed at every stage of food production, not just transportation. It would be absurd to only care about pollution emitted during transportation of food and show no concern for pollution at the stage of farm production. Over 80 percent of all carbon emissions of food occur at the farm and only 10 percent are emitted in transportation. By producing foods in the most efficient regions (e.g., pineapples in Thailand, lamb in New Zealand) the savings in energy at the farm level may lower the carbon footprint of food, even if that food requires greater food-miles. Or it may not—it just depends.

It is impossible to determine whether local food is truly more environmentally friendly, but we do know that fossil fuels are both emitters of carbon and a large component of a business' costs. If nonlocal lettuce is cheaper, that is a good sign that less fossil fuel was used, and thus a smaller carbon footprint results. Not even this rule is perfectly reliable though, as fossil fuels are just one of many costs involved in food production, and greenhouse gas emissions are not the only pollutant of concern. It is possible that a certain food could be cheaper and have a larger carbon footprint.

Does Buying Local Foods Stimulate the Local Economy?

It is true that spending dollars on imported food causes those dollars to leave the local economy, and that paying a local farmer $30 keeps that $30 (for a while) in the hands of your friends and neighbors. Buying local, then, seems to have an altruistic component, in that you are choosing to favor someone who lives close to you rather than a distant stranger. Many locavores thus argue that local foods are ethically superior because they provide economic support to those you know and favor. In addition, as that dollar paid to a local farmer circulates from one person to another in your area, your purchase of local foods acts as a local economic stimulus. One documentary on local foods has even claimed that spending one dollar on local food increases the region's total income by five or more dollars. If this were true every person in the modern world could become much richer by simply purchasing only local foods. If that sounds too good to be true, it is.

A publication by the Union of Concerned Scientists has suggested that each dollar spent on local food generates an extra $0.78 in income (in addition to the $1 spent) to the region—a more modest number, but still deceiving. This organization took the $0.78 number from studies published in the scientific literature—one study was even by our colleagues. Yet if you ask our colleagues, they will explain that the number is deceiving. It does not account for the fact that money spent on nonlocal food also generates additional income. Moreover, these studies do not account for how changes in spending patterns alter the imports and exports associated with the region. Put simply, the studies referenced cannot actually measure the net effect of buying more local foods. Let us explain.

This "economic stimulus" argument is an economic argument, one that does not depend on local foods being of superior quality to nonlocal foods, and so we assume throughout this section that the quality of both foods is identical. The local economic stimulus argument is indeed an economic proposition, but one with little economic theory or evidence to support it.

A core tenet of economics is that voluntary trade increases the wealth of all trading parties, regardless of whether it is two countries trading, two states, two counties, or even two people. Economists discuss wealth gains from trade like biologists discuss evolution: as a fact. Any time citizens would like to import and export to other regions but cannot, their aggregate wealth falls.

The local stimulus argument claims that, instead of freely trading with others, we should only exchange goods and services with those who live a few miles from us. Regardless of whether we are talking about restricting all trade, all trade in food, or even some trade in food, the locavore wants us to restrict trade, and both economic theory and empirical evidence say that this lowers the total wealth in all regions.

This is counterintuitive, as the idea of "keeping dollars local" just seems like it would be better for the local economy. Consider two arguments to the contrary. First, if the economic stimulus argument were true, then taking it to its logical extreme (which is one of the best ways to test logical propositions), it is better for Americans to only trade with people in their state—keeping the dollars local. It would also be better for Americans to trade only with people in their town—keeping their dollars local. It would also be better to trade only with people who live in their neighborhood—keeping dollars really local. Why not extend this to just one neighbor, or to deny yourself trade with anyone, that way your dollars never leave your pocket? Obviously, narrowing your opportunity for trade with others means you cannot have an iPad or any other advanced technology, and someone in North Dakota will never eat a pineapple. The local stimulus argument just isn't logical.

The second argument counters the "keep dollars local" claim. All imports into and exports from a region must be equal in value, over time. American exports equal American imports when measured in dollars, and exports from a small French town equal imports into that town. The philosopher and economist David Hume proved this in the eighteenth century, and his proof is supported by economists today (note: countries are reported to run trade deficits and surpluses only because the measured exports and imports don't count investments). This means that when you spend $100 on imported foods, that $100 does leave the town, but another $100 returns to the town in the form of exports. So whenever you import food, that money comes back to the economy, and all dollars essentially "stay local." If this were not true, any town would eventually collect all the money in the world or completely run out of money—something we never see happen. Take comfort, reader, that your money remains in the local economy regardless of whether you buy local or imported food.

This is an important issue because some influential organizations continue to push myths about local foods. The extension service at North Carolina State University urges its citizens to buy local because it will increase job opportunities and economic growth when in the university's agricultural economics department they teach the opposite.

Michael Pollan has suggested that we should force schools to acquire a portion of their foods within 100 miles, but local food is already an option for schools, and denying schools the option of importing foods simply makes it more difficult for them to access healthy foods within their budget constraints. If local foods were really cheaper and healthier, the schools would already be purchasing them. The secretary of agriculture under President Obama has even said that, in a perfect world, no region would import or export anything, a comment that could not be more contrary to basic economic principles. Anyone who believes that a "perfect world" would require Lockney, Texas (population: 2,056) to produce its own iPads or sugar has a very distorted grasp of economic principles.

> In a perfect world, everything that was sold, everything that was purchased and consumed would be local, so the economy would receive the benefit of that...
> —Tom Vilsack, secretary of agriculture, in "Tom Vilsack, The New Face of Agriculture," *Washington Post*, February 11, 2009, accessed September 3, 2010, at http://www.washingtonpost.com/wp-dyn/content/story/2009/02/10/ST2009021002624.html.

Is There Any Reason to Buy Local Other Than Food Quality?

In more recent years locavores have backed away from their claim that local food is better for the environment and local economy. They still make these assertions, but with less brio and more attention to other aspects of local foods. Even if spinach grown down the road is of the same quality but costs more, has a larger carbon footprint, and does not help the local economy, there is a reason we might want to encourage people to buy it. That reason has to do with our culture and attitudes towards food.

The locavore movement is not just about better shopping, but changing the food culture. They want us to think more about what we buy and its consequences, to take a greater interest in agriculture and food, to become involved in not just what we eat but what schoolchildren eat. They want us to mimic Transylvania by giving careful thought to the consequences of our food purchases. It is not their intention that we source all our food locally, but only a proportion of it. With this cultural change locavores suspect that the modern world would begin to eat a healthier diet. They may very well be right.

Consider Will Allen, a retired professional basketball player, who observed how some neighborhoods simply do not have access to affordable, fresh, and healthy foods. Doing his part to remedy this problem, he founded an organization that constructs greenhouses where organic vegetables are grown for the local community. This is not a business, but a nonprofit organization whose goal is to educate people about agriculture and healthy foods. In interviews with Allen and his fans they explain that they don't promote locally grown foods for the sake of local foods, but to help people who are unfamiliar with fresh vegetables to experience what cucumbers, basil, and "real" tomatoes taste like.

Urban people's interest in where their food comes from, and the quality of it—their worry about poisoned food, soil loss, toxicity, etc.—is a good thing. . . . If we stick only with the "local food" part of the movement, it's not going to amount to much. We've got to simultaneously talk about cultural change and land use more generally.

—Mary Berry, executive director of the Berry Center, "Mary Berry Is Fomenting an Agrarian Revolution," *Moyers & Company*, October 3, 2013, accessed October 6, 2013, at http://billmoyers.com/2013/10/03/mary-berry-is-fomenting-an-agrarian-revolution/.

What locavores are really trying to do is, in a way, to make Transylvanians out of us. If we can develop a similar love for agricultural landscapes and express an interest in how food is raised, it is thought we will eat better, love our food more, and become better stewards of our land. For those who identify with this sentiment, local food might be worth the higher price. It is, however, a poor reason to force people to buy local. Fortunately, most locavores are more interested in persuasion than force, and given the rising interest in farmers' markets, community supported agriculture, and the like, they have helped us become more conscious about the foods we eat.

8

CONTROVERSIES ABOUT LIVESTOCK

The Well-Being of Livestock Raised for Food

How Do You Define Animal Welfare?

Most people are omnivores, and because they also have empathy for farm animals, they want livestock to live a pleasant life—or at least not suffer. Our research has revealed that 31 percent of Americans believe that livestock have souls, and 64 percent believe that God wants humans to be good stewards of livestock. Only 28 percent say the feelings of animals are not important. Consumers express their altruism for animals in how they buy their food. Citizens express it in how they vote. Farmers and meat processors demonstrate it when they make large investments in better facilities and equipment to reduce animal stress, like those designed by Dr. Temple Grandin (as depicted in the HBO film that bears her name). Empathy is a constant concern for agricultural scientists as they research how to improve animal welfare while keeping food affordable, abundant, and safe.

Interest in animal welfare may be more pronounced today than ever before, but the interest has always been present. Each religion has its own particular way of viewing livestock, but they all respect the animal in some way. A theory for the development of the Jewish custom of consuming only kosher foods contends that the Jews sought to mimic the Garden of Eden, where humans relied solely on fruit, as eating both meat and vegetables involved death. Medieval Christianity around AD 1000 formed a movement called the Peace and Truth of God, which sought to protect livestock (as well as vulnerable people like orphans and widows) from violence by nobles. Hinduism developed a particular reverence for the cow, and some Buddhist sects believe animals harbor the souls of past and future humans, and thus should be treated with the same compassion.

Starting in the nineteenth century, the concern for animals led to the formation of animal advocacy groups. Animal protection agencies were first formed in Britain, which inspired Americans to form the Society for the Prevention of Cruelty to Animals (ASPCA) in 1866. This organization soon persuaded politicians to pass laws regarding how livestock are transported by rail. The laws set a precedent followed by animal advocates to this day.

The nineteenth century also introduced a new moral philosophy, utilitarianism, which revolutionized how the educated think about animals. Most readers will attest that it was a vast improvement over the old one. In the seventeenth century the famed philosopher Descartes had argued that animals were mere machines, devoid of emotion. His entourage would beat animals in public and mock those who expressed empathy. Fast-forward to 1823 and philosopher Jeremy Bentham is sketching out his concept of utilitarianism, arguing that the suffering of animals may indeed be of similar moral interest as the suffering of humans. It would take more than a century for this philosophy to affect how people think about animals, but its impact was eventually manifested in the writings of animal advocates and agricultural scientists alike.

> The day may come, when the rest of the animal creation may acquire those rights which never could have been withholden from them but by the hand of tyranny.... But a full-grown horse or dog is beyond comparison a more

rational, as well as a more conversable animal, than an infant of a day, or a week, or even a month, old. But suppose the case were otherwise, what would it avail?...The question is not, Can they reason? Nor Can they talk? But, Can they suffer?

—Jeremy Bentham, *Introduction to the Principles of Morals and Legislation*, 2nd ed. (London: Pickering, 1823), chapter 17.

Bentham's utilitarianism philosophy, espousing the idea that public policy should be designed to maximize total happiness and minimize total suffering, would become especially important to Peter Singer. In his book *Animal Liberation*, Singer used utilitarianism to argue that most livestock production was immoral and that consumers should cease to eat such foods. It was this book, along with Ruth Harrison's *Animal Machines*, that launched livestock welfare controversies in the 1960s and 1970s, which are still debated today and discussed in this chapter.

Animal Liberation did not necessarily call for a vegan diet, though. Singer's particular version of utilitarianism suggested that the raising of livestock for food can be ethical if the animals are treated humanely. In a later book with Jim Mason, *The Way We Eat*, Singer takes readers on a tour of various farms to help them distinguish between humane and inhumane food. So even though Singer is often cast as an extreme animal rightist, his general philosophy is in tune with the average American: that livestock raised for food should be treated humanely. Where Singer differs from the average American is in his view of what constitutes "humane."

Why is it that one person can deem a particular style of farming humane while another person dissents? Part of the disagreements arise because no one really knows what an animal is feeling. It must be inferred based on common sense, biological measurements, and animal behavior—but widely differing conclusions emanate from those inferences. Three general schools have emerged on how animal welfare is measured. These are the (1) function-based, (2) feeling-based, and (3) nature-based schools. These different schools have provided a framework within which the public debate on the treatment of domesticated animals has taken place since the early twentieth century. The schools do not compete for legitimacy, as all are considered valid ways of measuring animal welfare. However, there is overlap between schools, and many times, the importance given to each school will vary from person to person according to philosophies, experiences, culture, and societal influences. Ultimately, animal welfare is best served when advocates combine the most rational features of each school, particularly where all three intersect.

It can be difficult for the average person to understand the modern livestock farm, but those who have pets understand more than they think, so we will use the analogy of caring for a dog to help readers understand how and why livestock industries raise livestock the way they do.

Humans care deeply for their pets. Many smokers say they are more likely to quit smoking for their dog's health than their own. Lawyers have argued that pets be recognized as family in courts. Some Christians even baptize their pets. People certainly don't have these feelings about cows, chickens, or hogs, but in some aspects it seems as if they do.

Most dog owners demonstrate their love by purchasing dog food scientifically tailored to their dog's breed, age, and size, and they take their dog regularly to the veterinarian. These owners want their dogs to function well biologically; this is an example of the *function*-based school of animal welfare.

Likewise, farmers also keep their animals biologically fit so that they are healthy enough to grow and reproduce. In recent years, pet foods have become increasingly sophisticated and targeted. One brand trumpets its sophistication with the name "Science Diet." The process behind livestock feed is arguably more scientific. Visit a dairy farm to see how the cows' feed is formulated. As a supplement to hay and silage, the farmer may add soybean meal, yeast, hominy corn, rendered blood, minerals, and other things that cannot be consumed by humans and would otherwise end up in a landfill. The precise amount of each ingredient is calculated by expert nutritionists who use computers to account for the animal's every nutrient need, in ways far more deliberate and scientific than what we feed our pets—or even what we eat ourselves. One of the authors used to work on a dairy that attached collars to the cows so that, when the animal approached the feeding stall, it activated a computerized feeder which delivered its precise dietary needs. Every time the cow ate the event was recorded, so that the farmer could be quickly alerted when a cow had not eaten (and was therefore probably sick and needed immediate attention).

One of the greatest dangers to a sow (pregnant pig) is other sows. When many sows are given access to the same food, water, and space, they will fight over those resources, causing both physical and mental harm. The dominant sows will eat too much, the subordinate sows too little. Like cows, hogs can be given collars or ear tags, and an automated feeding stall that allows only one animal to enter at a time. This allows the sow to eat in peace, while making sure it is protected from other aggressors and doesn't eat too much or too little.

Welfare-conscious improvements in poultry feeding strategies have also been developed. Chickens used to be denied food from five to fourteen days at a time to induce synchronized molting (the natural feather shedding process used to rejuvenate hen egg production) in an entire flock. Since the 1980s, many concerns arose regarding the welfare of hens being starved in order to artificially promote molting, which resulted in a variety of alternative methods. Today, special feeds and diets allow the hen to molt without experiencing hunger.

Think back to dogs and their owners. Dogs owners protect their pets from excessive heat and cold by housing them indoors or in doghouses. To prevent infestations of worms, fleas, and ticks, the dogs' living areas are kept clean, and they are given medication to ward off parasites. Likewise, most hogs and chickens are now raised indoors, not only to keep them at a comfortable temperature and protect them from predators, but to reduce disease. Cows are regularly given medicine that wards off flies, fleas, lice, and ticks, and bedding for dairy cows and chickens are routinely cleaned.

If you have not spent much time on a farm, throw away every notion you have about it being old-fashioned. The modern farm is a highly specialized, technical, and scientific business. Some cows today even have stomach capsules that record the animals' temperature, alerting the farmer if it is running a fever. On her smartphone the farmer can install the Thermal Aid app, which collects a variety of weather data and alerts the farmer when heat stress in cattle is likely. Because animals generally perform well when their biological needs are met, the livestock industry has become expert at meeting the function-based needs of the animal.

Yet meeting the function-based needs of the animal is not sufficient for everyone. Most dog owners subscribe to the *feeling-based* school of animal welfare. They will swear they can tell if their dog is sad or scared, which makes the owner respond with love and comfort. Though she might not feel like it at the time, an owner may take her best friend for a walk if the dog whines at the front door, or stop to play if the dog seems bored.

One might be tempted to say that the livestock industry manages too many animals to account for their feelings, and that feelings cannot be scientifically measured, anyway. While it is true that farmers cannot get to know every pig and cow, they do care about animal feelings, and scientific tools for indirectly measuring how animals perceive their environment do exist. For example, mental states of stress in animals are correlated with elevated levels of certain hormones. A stressed sheep, for instance, will display higher cortisol concentrations when it is separated from the flock.

Building shelter for animals is expensive, so a farmer will want to keep as many hogs under one roof as possible, but not so many that the stress of the crowd is detrimental. In experiments, animal scientists have placed groups of hogs at different stocking densities (different square meters per pig) and measured cortisol levels of each group, thereby allowing one to determine a density that keeps both animal stress and farmers' costs low.

Another way of inferring an animal's mental state is to simply let the animal reveal its preferences by presenting it with choices. These choice experiments even teach animals to pay a "price" to receive something; instead of paying with money, the animals pay by performing a physical action. The more times that action must be performed, the higher the price it pays. So not only do we measure whether livestock prefers one thing or another, but also the maximum price they will pay for each (i.e., a measure of their motivation to obtain a resource). For example, it is known that hens truly desire nests for laying eggs because they are willing to squeeze through a very small hole to reach one. Even if you raise the price by making the hole smaller, they squeeze through whenever it is physically possible. Hogs want both to eat and to socialize with other animals; we know this because they will press a lever multiple times for both. They value food more than socialization, though, because they will press a lever more times for food.

Scientists study these animal preferences using experiments and mathematical models, the same models economists use to capture human preferences. Some economists have gone so far as to measure "total happiness" and "total suffering" by humans *and* animals in one utilitarian function, thereby achieving what Jeremy Bentham and Peter Singer have long advocated: taking animal feelings into account just as one accounts for human feelings. So, like dog owners, the livestock industry takes animal feelings seriously—just in a more technical way.

Finally there is the *nature-based* school of animal welfare, which simply says that animals are content and comfortable when they are allowed to express their natural instincts and live in natural environments, and discontent, stressed, and uncomfortable when they cannot. Every dog owner knows the most important part of a walk is when the dog smells the urine and feces of other dogs. It doesn't accomplish anything practical for the dog, but is still an essential part of a happy dog's life. They also like to play tug-of-war, chase squirrels, and protect their master from the UPS man, all because these were once essential behaviors of their wild ancestors.

Like dogs, livestock still desire to engage in certain natural behaviors even if they have no tangible benefit. Chickens like to scratch in the sand even if their feed trough is full, and they like to bathe in dust even if there are no parasites. Hogs absolutely love to dig, explore, and wallow in the mud. Cows prefer shade during hot summers and to live in herds.

Farmers acknowledge the importance of these natural behaviors and, when it is economically feasible, are happy to provide outlets for all of them. For example, instead of a barren cage, laying hens can be housed in "enriched" colony cages, which contain perches, sand for scratching and/or dust bathing, and private nest boxes when the consumers are willing to pay a premium to cover the additional cost of production. Some farmers allow their hogs outdoor access or, if indoors, sawdust for them to excavate. Farmers rarely house adult cows individually, allowing them to live in herds, and when available, to have access to shade and pasture.

What Is the Animal Welfare Controversy?

Even the most loving dog owners cannot provide their pet constant bliss. Sometimes there are trade-offs between two things that make a dog happy. Immediately before an owner leaves for work, the dog begs to play in the fenced-in backyard, but outside it would have to stay in the cold all day. Thus, denying the dog the ability to play for a few minutes outside prevents it from being cold all day. Or the dog is getting fat and its owner imposes a dreaded diet. Though it wants to eat so much more, the owner knows that denying it unlimited food is in its long-term best interest.

For similar reasons, farmers must sometimes sacrifice one aspect of animal welfare to provide another. Hogs are usually raised on concrete where they are denied the ability to explore, rest comfortably (though hogs often prefer concrete in the summer), and dig. Yet being on concrete makes for more sterile housing, thereby improving the health of the animal (and the safety of the meat). Laying hens are sometimes placed in a small, barren cage with a few other birds. This makes for an unenriched environment, but if allowed to roam free in the barn with thousands of other hens they will injure one another, causing pain and resulting in higher death losses. The hens would also like to go outside and hunt for insects, but free-range farms can have very high mortality rates (up to 25 percent, compared to 3 percent in cage systems). It is true that cattle in feedlots are not allowed to graze in pastures, but it is also true that they are given an alternative feed that they crave even more than grass and are given individualized attention more frequently by animal managers.

There are times when you must choose between your happiness and the dog's—sometimes you choose your own. It wants to go on a walk, but you are tired and your favorite show just came on. It misses going into the backyard whenever it wants, but you can't afford to repair the fence right now. There are even cases when a pet is euthanized because the surgery to keep it alive without pain is too expensive.

Similarly, most consumers will simply not pay the premium necessary to provide all animal needs. If farmers must sell food products at the lowest prices, then farms must operate efficiently, and this requires them to sacrifice some elements of animal welfare to keep food affordable. For example, there are some hog farms that provide most of what hogs need to live a pleasant life, including a sanitary environment, protection from aggressive hogs, space to explore, mulch to dig in, and more. Few hog farms like this exist, though, because it costs up to 30 percent more to raise hogs, and few consumers are willing to pay this premium. There is a movement in the United States led by both the United Egg Producers and the Humane Society of the United States to convert all egg farms currently using small, barren cages to enriched cages. They wish to do this by lobbying for federal legislation requiring enriched cages and standardized cage sizes across the United States. There is little doubt that hens are able to behave more naturally in enriched cages, but it would raise egg production costs by around 12 percent, and if consumers would voluntarily pay this premium, no legislation would be necessary, as farmers would voluntarily adopt them. However much consumers, farmers, and pet owners want animals to experience superior states of welfare, there is a limit to what they will pay to achieve it.

Husbandry Practices and Medical Procedures

There is little controversy about what type of environment animals need to achieve optimal welfare. Controversy exists because few people are willing to pay the price necessary to provide all these needs, so the argument revolves around which needs should be sacrificed. To provide some examples, consider table 8.1, listing various medical procedures that, on the one hand involve pain and stress, but on the other hand provide benefits to both the animals and consumers.

Readers are probably familiar with the black-and-white dairy cows on Chick-fil-A commercials. Next time you see the commercials notice that the cows do not have horns. This is not because that breed is hornless, but because the horns were surgically removed by the farmer. While dehorning is certainly painful and is often performed without anesthetics, it prevents the cows from injuring each other and farmworkers. Male pigs are castrated because boar meat is inferior, and

Table 8.1 Examples of Routine On-Farm Management Procedures Conducted on Livestock

Species	Management practice	What the practice involves	When this practice is typically done	Why this practice is done
Cattle	Early removal/wean from mother cow in dairy animals	Separation of newborn calf from its mother	Immediately to 48 hours after birth	Allow for milk production from cow to enter the food supply; calves are provided with a milk replacer.
	Castration of males	Removal of testicles from male cattle not used for breeding purposes	Dairy calves: ~3–4 weeks of age Beef calves: ~10–13 months of age	Male cattle are safer and easier to handle, and castration prevents unwanted breeding.
	Tail docking of female dairy cows	Removal of the tip of the tail or shortening of the tail	A few weeks after being weaned off milk or a few weeks before giving birth	Make access to the cow's udder easier during milking
	Disbudding/dehorning	Prevention of horn growth or horn removal/amputation	Dairy: 1–6 weeks of age Beef: 2–12 months of age	Protect cattle and farm workers from injury
	Animal identification	Pierce the ears with ear tags and/or marking the skin by hot iron or freeze branding	Upon birth or arrival to a farm	Provide means of identifying individual animals.

	Procedure	Description	Age	Purpose
Pigs	Tail docking	Removal of 1/3 to 1/2 of the tail	3 days–8 weeks of age	Prevent tail biting by other pigs.
	Teeth clipping	Clip or grind off the canine teeth of piglets	3–8 days of age	Reduce injury to littermates or to the udder during nursing by piglets
	Castration of males	Removal of testicles from piglets not used for breeding purposes	3 days–8 weeks of age	Prevent boar taint (in meat) and aggressive behavior problems
	Animal Identification	Cutting a small triangular section out of the edge of a pig's ear	3–8 days of age	Provides a permanent, inexpensive identification system to individually identify animals
Poultry (chickens and turkeys)	Beak trimming	Beaks are trimmed and no more than 1/2 and 1/3 of the upper and lower beak is removed, respectively.	1–10 days after hatching or 18–16 weeks of age	Prevent cannibalism and pecking between birds that may lead to skin injuries or feather loss
	Claw removal of males	Removal of the last joint of the inside toes of male breeding birds	1–3 days of age	Prevent injuries to hens during mating

because it reduces aggression between pigs. The beak of a chicken is a weapon, and adult hens can be surprisingly cruel, so their beaks are trimmed at a young age. These are all examples of trading one aspect of animal welfare for another, and not everyone agrees on whether the trade-off is ethical.

Some controversies regard whether a procedure should be performed. For example, some dairy farmers used to routinely dock the tails of their cows. Recent research suggests the benefits are small or nonexistent, and so most farmers now leave cows' tails intact, and docking is even banned in numerous states in the United States (California, Rhode Island, New Jersey, and soon Ohio [2018]) and Europe (Denmark, Germany, Scotland, Sweden, the United Kingdom, and some Australian states). The European Union is attempting to eliminate all castration of male pigs by 2018. Other times the debate is not whether, but how a procedure is performed, like whether castration is accompanied by anesthesia, a practice that is mandatory—though not always followed—in the European Union (if the piglet is older than six days).

Housing

Although controversies about the severity of management procedures will continue to exist, the most consequential debates concern livestock housing, and this is probably because consumers and citizens generally dislike housing animals in small, barren cages. In the United States, for example, most egg production takes place in battery cages where a few hens are placed in a barren wire cage, as shown in Table 8.2. The wire allows manure to drop through the floor onto a conveyor belt, keeping the cage sanitary, and with few hens in each cage, aggression is not much of a problem. These cages are now banned in the European Union and three US states. Will these bans improve hen welfare? It depends on what replaces the cages. If the barren cages are replaced with larger enriched colony cages, many agree that the hens' behaviors and ability

Table 8.2 Controversial Housing Systems

Name of housing system	Species housed	When used in the animal's life	Why the housing system is used		Cons
			Pros		
Conventional or battery cage	Egg-laying hens	All their life (~18–24 months)	Sanitary		Little room to move.
			Protection from predators.		Restricts natural behaviors like dust-bathing, walking, foraging, nesting, wing flapping, stretching, body shakes, tail wagging, and roosting.
			Reduced death and injurious pecking from other hens.		
			Easier to observe birds and provide medical treatment.		
			Economically efficient.		Reduced bone strength.
Gestation crate	Pregnant sow	During pregnancy (~115 days)	Reduces aggression and injury between sows.		Sow can neither walk nor turn around.
			Sows do not need to compete for food.		No comfortable area to rest.
			Protection from heat and cold.		Boredom can cause stereotypic behavior (odd repetitive behavior).
			Easier to provide medical treatment.		
			Economically efficient.		
Farrowing crate	Sows	During birth and nursing (~20–30 days)	Protects piglets from being crushed by mother.		Sow can neither walk nor turn around.
			Gives piglets constant access to sow.		No comfortable area to rest.
			Provides piglets choice between cooler and warmer areas.		Boredom can cause stereotypic behavior (odd repetitive behavior).
			Flooring is dry and sanitary.		
			Easier to assist sows in birthing.		
			Sows do not have to compete for food.		
			Economically efficient.		

to express behaviors will be improved, but there is less consensus on hen health if the cages are replaced with cage-free facilities (i.e., aviaries or free-range systems). Some believe that, although mortality rates are higher in a cage-free system, this is a cost worth paying if the hens are given ample room to move around, explore, and express their normal behaviors. Others believe animal welfare is lower in a cage-free facility, due to increased predation, cannibalism, hen piling and smothering, feather pecking, exposure to and spread of parasites and diseases, and mortality, and that these factors should not be ignored when evaluating welfare and a housing system.

Regarding what hens "feel" about their environments, this disagreement is almost impossible to reconcile because it is impossible to measure whether hens are truly "happier" or emotionally distressed in any one system. However, using behavior experiments comparing normal ancestral behavior patterns to domesticated hens and studying preference tests (for resources like perches, private nest boxes, increased space, etc.) can help provide insight into the extent that hens desire or dislike environmental features (i.e., their motivation to live in a particular environment or obtain a specific resource). These behavioral scientific tools can indirectly provide us the information needed to better understand the feelings hens may have about their environmental conditions. With many different ways of looking at welfare, it is clear that assessing hen welfare can be quite difficult and multifactorial.

Another pet analogy that one of the authors uses to illustrate how people have different priorities regarding welfare was described by the welfare and behavior research specialist Dr. Joy Mench of the University of California, Davis. This analogy describes an indoor cat sitting at a window, longing to go outside and explore, hunt, and protect its territory. Some cat owners (like the author herself) recognize that their cat's desire to act on its natural instincts is important to its mental well-being. But is allowing the cat to have outdoor access worth endangering the cat by exposing it to vehicles, predators, other cats, parasites, diseases, and weather extremes? This depends on what the owner values more: the ability of the cat to fulfill its desire to express certain behaviors outdoors or maintaining the safety of the cat from outdoor hazards. Keeping the cat indoors minimizes the risk of injury, illness, or death by outdoor factors, but some owners are willing to let their cat face those risks in order to perform behaviors driven by its natural instincts. Both types of owners have their cat's "best interest" in mind; they just prioritize those interests differently.

Hogs are perhaps one of the most difficult livestock species to manage. The fences used for horses and cattle cannot contain them, and they can turn a flat, verdant field into a World War I no man's land in a short period of time. Sows are particularly stubborn and sometimes require extra time and effort to move them from point A to point B. Providing medicine, assisting in birthing, and artificial insemination (the norm) is much easier when sows are placed into steel stalls with a slatted floor and sides so narrow the sow cannot turn around. These are called gestation stalls when sows are pregnant and not yet nursing piglets. Farrowing stalls also are narrow and prevent the sow from turning around, but provide piglets protection and room for movement to nurse. The stalls allow the farmer to feed and treat the sow individually, and it protects her from other sows. While they provide certain benefits to the sow and farmer, the inability to turn around or walk obviously frustrates the sow. If farmers do not use gestation stalls, most will use a group-pen, which is simply a barren pen with a few too many sows. Each animal has the ability to move around, but is now at the mercy of sharing resources with dominant sows. Are they better off in the group-pen? That depends on whether the injuries and competition for food outweigh the benefits of greater mobility.

So once again, it will depend on what replaces the gestation stall and how that alternative better protects a sow from injury, behavioral problems, health impairments, mortality, and other environmental conditions that negatively impact animals. Sometimes it is a lateral move to change from one system to another in terms of welfare indicators, because current alternatives still present pros *and* cons to sow welfare; but now there's the added expense of installing an entirely new system and training workers to learn how to manage the new system. Some believe such a change may be worth the trade-offs, whereas others do not.

Consider a specific case one of the authors researched as part of her graduate studies, involving male dairy cattle raised for beef. Since the calf is taken from its mother at birth, the farmer must perform all the duties of the mother. Calves are especially vulnerable to sickness at this age, so a sanitary environment is imperative. One housing option is a hutch with slatted floors, where the manure falls below the animal and separates the animal from its excrements, reducing parasites and disease, but providing only hard surfaces for the calf to rest. An alternative is a hutch with bedding, like straw or sawdust, but no separation of the calf and its excrements. The calf may have a preference for the hutch with bedding (depending on the weather). However, it is unaware that the pen with bedding will put it into greater contact with its own feces and bacteria, increasing the chance of sickness, so we can't simply ask the calf which pen is better.

The author's research allowed her to measure the extent to which bedding might increase sickness. For instance, it quantified the differences in airborne bacteria concentrations, finding that bedding resulted in more than twice the concentrations of airborne bacteria than slatted floors. However, the research could not say that one system was better than the other for the animal overall, because both options had both pros *and* cons for the well-being of the calf. The research can only point out where areas of concern may exist, so that current and future systems can prevent conditions that can impede the welfare of animals.

Periodically, such research provides unambiguous progress, and this did happen in the aforementioned study. In addition to comparing bedding versus slatted floors, the author evaluated a third system, which was bedding with an additive that reduced the bedding's pH to improve the sanitation of the bedding. Results from this study and previous studies found that this additive did indeed reduce bacterial concentrations and even fly survival, providing more insight on strategies that can improve the bedding conditions of calf housing.

Handling

One cannot discuss animal welfare without addressing how livestock are handled, restrained, managed, and transported. One of the most famous animal scientists is Dr. Temple Grandin, and anyone who watched the HBO film about her life (titled *Temple Grandin*) knows that she designed facilities and handling methods that reduce stress in cattle while also making handling easier for workers. More than any other she understands the mind of an animal, and her keen insights have revolutionized how livestock are treated on farms and in slaughtering facilities. For instance, animal science students today learn about the "flight zones" of different livestock species, which allows one to direct animals to their intended destination with less stress and injury for both humans and animals. Compared to thirty years ago, livestock handlers today use electric prods less often, and are gentler, quieter, and more aware of how the animal is thinking. Grandin didn't just change the equipment that industry uses, but the cultural norms about how animals perceive their environments and are handled by humans.

With other animal scientists and the livestock industry, Grandin has developed objective auditing procedures that companies can use to identify problem areas, like slippery floors. This makes it easier for a company's overall welfare standards to be audited, and auditing is important if the company wishes to assure customers that its methods are humane.

There are occasions when standard handling techniques appear questionable or cruel on video but in reality are more nuanced. For instance, scenes from the documentary *Samsara* (http://vimeo.com/73234721) shows a machine (referred to as a mechanical chicken catching system) traveling on the edge of a dense crowd of birds within a barn, catching birds with soft, rotating fingers, and moving them onto a conveyor belt, where humans pick them up and place them in a bin to be transported.

The catching machine is actually considered to be quite humane, both for the birds and workers. Without the machine, intensive and strenuous human labor is needed, where workers enter large flocks and pick the birds up by hand, holding them upside down with three or four other birds and then placing them in a cage for transport. The birds experience stress when being caught, regardless of which method is used, but there is evidence that stress and injury rates are lower with the machine. Perhaps future research will find newer technologies to advance and improve handling methods, but the important thing is to keep asking how animal welfare might be improved and conduct scientific research to parse the good ideas from the bad.

Even a casual perusing of YouTube using the search terms "undercover + investigation + farm" will return a number of videos showing pigs, cattle, and chickens handled cruelly. Some are too difficult for most people to watch, but often, many of the videos are interspersed with both cruel treatment and acceptable day-to-day management practices. When such videos are viewed by the layperson, all practices presented on the video are lumped into a single "inhumane" category that further confuses those far removed from agriculture and obscures the discussion of livestock welfare. While depictions of cruel treatment are not representative of most farms and are misleading to viewers, they sometimes show what is possible when humans do not abide by accepted norms of animal treatment. Every new depiction of cruelty posted online is not only publicly condemned by animal advocates and consumers, but livestock industry groups as well. This does not mean that the livestock industry and animal advocacy organizations are on the same page. They still disagree on how and whether live-stock should be raised for food, but the difference is that, today, almost everyone expresses a commitment to animal welfare.

Hopefully, these examples demonstrate that the contro-versy involves far more than being "kind" to farm animals. They also concern how one should go about being kind. Two individuals may have equal empathy for animals but disagree on how the animals should be raised (i.e., their priorities differ across the three general schools of welfare: function-, feeling-, and nature-based schools). While agricultural scientists are not in a position to tell society how much compassion livestock should be given, they can play—and eagerly wish to play—a productive role in helping society understand how to achieve the best welfare of all animals in human care.

Scientists do not make many decisions about how animals are treated, though. These decisions are made by the interac-tions between agricultural industries, their customers, the public, special interest groups, and policy. It is to these interac-tions we now turn.

How Is Farm Animal Welfare Regulated?

Being US researchers, we will approach this question largely in terms of US regulation, but the reader should be made aware that the European Union has taken farm animal welfare more seriously than any another other region. It continually prescribes new minimum welfare standards that all EU coun-tries are expected to meet, and these standards usually place more emphasis on the feeling-based and nature-based schools of animal welfare, compared to the United States.

Regulation of livestock production in the United States tends to take place at a state level. The federal Animal Welfare Act specifically excludes farm animals, but there are federal laws to prohibit inhumane slaughter and regulate how live-stock are transported, and one federal law requires the sec-retary of agriculture to study a variety of livestock issues, including animal welfare. There are no federal laws about how livestock should be treated while they are on the farm.

Some readers may have first come across some of the hog welfare controversies when they watched HBO's documentary *Death on a Factory Farm*, where an undercover animal rights activist covertly filmed a hog farm where he worked. Some of the scenes depicted a typical hog farm, with sows in gestation crates and pigs on concrete slabs, and these are conditions that some animal scientists and veterinarians find acceptable. Most of the film concerned acts that the vast majority of farm-ers do not condone, like providing inadequate feed, allowing cannibalism, and "euthanizing" sows by hanging (hanging was still legal in Ohio, though). The farm operators were charged on eight counts of violating Ohio's anticruelty laws. Such counts are difficult to prosecute, because the laws often exclude livestock from certain requirements. For instance, Ohio laws state that no person should "keep animals other than cattle, poultry, or fowl, swine, sheep, or goats in an enclo-sure without wholesome exercise and change of air." If the farmers had kept dogs under the conditions described above, they would have been found guilty, but because the animals were pigs, they were not.

Until recently US farmers were allowed to house their live-stock however they pleased, but in the last decade a number of changes in state law have forced egg producers in three states to seek alternatives to the battery cage and hog farmers in eight states to care for sows without gestation stalls. Citizens in the states of California, Arizona, and Florida voted on initiatives to ban one or both of these housing systems. The California initiative, referred to as Prop 2, receive considerable attention, so much that Oprah Winfrey devoted an entire episode of her show to it, even a bringing gestation stall and battery cage to her studio for the audience to see for themselves. The states of

Oregon, Colorado, Maine, Michigan, and Ohio banned at least one of these systems through legislation.

Consider the wording of California's Prop 2, which reads as follows.

> A person shall not tether or confine any covered animal, on a farm, for all or the majority of any day, in a manner that prevents such animals from (a) lying down, standing up, and fully extending his or her limbs; and (b) turning around freely.
>
> —Proposition 2, State of California, 2008, accessed November 25, 2013, at http://ag.ca.gov/cms_pdfs/ initiatives/2007-08-09_07-0041_Initiative.pdf.

At the time it was passed, many farmers were not sure how to comply with the proposition since specifics on acceptable housing methods were not established by the proposition nor the Humane Society of the United States (HSUS). Notice the proposition does not ban cages. Are cage-free production methods the only acceptable methods? What about caged systems (such as enriched colony cages) where birds *are* able to lie down, stand up, extend their limbs, and turn around? This was not what the HSUS had in mind, though, and the disagreement led to a lawsuit by the Association of California Egg Farmers (and others) requesting clarification about exactly what Prop 2 means in regard to acceptable housing systems. There was also the possibility that California might continue to import eggs from caged systems in other states (as it had regularly imported eggs prior to the passing of Prop 2), but that was prohibited by later legislation.

HSUS and animal scientists in the United States never agreed about cage-free egg systems. Although HSUS believed it to be the most humane method of egg production, animal scientists could only agree that cage-free systems allowed greater movement and natural behaviors at the expense of greater mortality and injury rates. It is nearly impossible to find any animal scientist in the United States who expresses with confidence that cage-free systems are unambiguously better for hens than the cage system, or, for that matter, that cage systems are better than cage-free. This is because there are trade-offs (pros *and* cons) in every housing system.

Then came the unexpected. Just as the debate between cage and cage-free egg production looked as if it were about to become bitter and prolonged, the United Egg Producers (largest egg producer group in the United States) and HSUS reached an agreement on how hens should be raised. In 2011 they both agreed to jointly pursue federal legislation requiring enriched colony cages (among other farming standards), where hens are given sufficient space to comply with Prop 2, plus be allowed perches, scratching areas, and private nest boxes. Moreover, the animal scientists who advise the United Egg Producers supported the plan, and enriched cages are generally regarded by most animal scientists to provide for better hen behaviors than cage systems.

Why did these two opponents reconcile? The HSUS probably believed it would benefit laying hens, although HSUS has historically campaigned for cage-free egg production. It likely knew it would also set a precedent by which it could lobby for more regulations at the federal level. The egg producers were perhaps motivated by the fact that state-specific laws on caged eggs are so variable and this legislative agreement could halt costly battles across state lines. Likewise, if producers in certain states were to incur higher production costs, they surely would want producers in other states to do the same—thus the push for nationwide uniformity. Some have even suggested that higher animal welfare standards allow the egg industry to collectively reduce egg production and boost prices, but there is no public evidence to support this claim.

A mutually agreeable solution was reached, and the problem was solved, it seemed. Senator Diane Feinstein (D-CA) introduced the legislation in May 2012, but then other livestock groups became concerned that this would set a bad precedent, and might lead to federal legislation affecting other livestock farmers. Just when egg producers and animal advocacy organizations seem to have settled on an agreement, other livestock producers have now entered the debate, because the manner in which the egg controversy is settled impacts how, say, the pork controversy is settled. As the time of this writing, the egg controversy is still unresolved.

> This HSUS-backed legislation would set a dangerous precedent that could let Washington bureaucrats dictate how livestock and poultry producers raise and care for their animals.... We don't need or want the federal government and HSUS telling us how to do our jobs.
>
> —Doug Wolf, president of the National Pork Producer Council, "Livestock Groups Equate HSUS / UEP Bill to Government Takeover of Farms," Agri-Pulse.com, January 24, 2013, accessed November 25, 2013, at http://www.agri-pulse.com/HSUS_UEP_ legislation_012312.asp.

Producers in other industries may oppose federal legislation on egg production because they believe that mandating a "one size fits all" federal bill would (1) take away producers' freedom to operate in manners they see fit for the best of their animals, (2) make it challenging to respond to consumer demands and choices, (3) increase food prices, (4) negatively impact niche markets and small-scale farmers, and (5) redirect budgeted funds from enhancing food safety and US competitiveness to regulating on-farm practices for reasons other than public and animal health. Thus, industries other than egg production fear this agreement would set a "dangerous precedent" for the future of their own industries; and not only for the reasons mentioned, but also for the fear that animal rights groups, like HSUS and PETA, could dictate on-farm practices when the mission of such groups is to abolish the use of animals by humans, which greatly differs from the views of those in food production.

To see why pork producers might oppose federal legislation on egg production, consider two important facts. First, although gestation stalls have been banned in eight states (or, more accurately, will be banned at a precise future date), those states raise relatively few hogs. Second, most pork production takes place in states that are protected from state-level initiatives to ban gestation stalls. The states with the highest hog populations are also ones in which state-level initiatives are either not allowed, or the requirements to get initiatives on the ballot are so stringent that an initiative to ban gestation stalls seems almost impossible. Gestation stalls in the United States are thus safe from state-level legislation, but producers don't want the possibility of federal regulations either.

The pork industry is under some pressure from retailers, though. Companies like Kroger, Subway, McDonald's, Denny's, Target, Sysco, Oscar Mayer, and Conagra have announced intentions to source pork produced without gestation stalls. Gestation stalls do lower production costs, but converting to group pens only raises the cost of producing retail pork by about 2 percent. For these reasons, some in the pork industry are making moves towards a voluntary switch from gestation stalls to group pens, a move Smithfield Foods (the largest pork producer in the world) announced it was making years ago. This doesn't mean gestation stalls will soon be a historical relic, for animal scientists are far from agreed that group pens are better. Some, like Dr. Janeen Salak-Johnson of the University of Illinois, have been especially vocal in lamenting that decisions about how sows are raised are beginning to be made not by farmers, animal scientists, and veterinarians, but retailers and restaurants. What does seem certain is that some portion of the pork industry is converting to group-pens due to pressure from its buyers. Whether this transition evolves to an industry-wide phenomenon remains unclear.

Growth Hormones in Livestock Agriculture

Beef Cattle

One of the authors once took a college class giving him hands-on experience caring for newborn calves. Every morning for two weeks he would drive around a pasture looking for calves born the previous night. Once a newborn was spotted, he would castrate the animal if it was a male, attach an identification tag in its ear, and inject into the calf's ear a small pellet containing synthetic growth hormones (often, estrogen). With this hormone the calf would be healthier and grow faster. The hormone's impact on cattle growth is so large that ranchers receives between $5 and $10 for every $1 they spend on the hormone.

Many years later the author was talking to a cattle producer who remarked that the use of growth hormones was causing young women to mature faster. The author laughed, as the statement seemed so outrageous that he assumed the farmer was joking—but he was not.

If you research the issue (but not a lot) you can see where the farmer is coming from. You are putting growth hormones into cattle, so it is only logical that the hormones might be present in the beef, possibly causing changes in the person eating the beef. The rumor that this causes early puberty has some basis in fact. One of the worst scandals to hit the dairy industry was in 1974, when cattle feed was accidentally laced with a flame retardant called PBB and fed to thousands of cows in Michigan. Before the accident could be discovered, most Michigan citizens had drunk milk from those cows. The key detail is that this PBB mimics estrogen, and there is some evidence that pregnant mothers who drank more of this milk gave birth to girls that matured earlier.

Add to this true story the fact that synthetic growth hormones have been used on a large scale since the 1960s, and that young people started to mature younger in the 1970s, and that explains the fear that growth hormones in cattle production are causing children to mature earlier.

Dig deeper, though, and it is clear that hormone use in beef production bears no resemblance to the PBB scandal. First, children may not even be maturing earlier. Studies that suggest they do tend to rely on subjective judgments about the size of breasts and testicles, but when you use an objective measure like the age at which a girl first menstruates, a trend towards earlier maturity isn't found.

Even if kids are maturing earlier, it cannot be because of synthetic growth hormones in beef cattle. It is true that cattle given estrogen will have more of the hormone in their meat: an additional 0.4 nanograms of estrogen for every 4 ounces of beef, relative to cattle not receiving estrogen. Compare that to four ounces of raw cabbage, which contains 2,700 nanograms, a soy latte (with one ounce of soy milk) which contains 30,000 nanograms, or three ounces of soybean oil, which contains 168,000,000! One birth control pill contains 25,000 nanograms, and the average prepubertal girl will have 54,000 nanograms of estrogen in her body every day.

All growth hormones given to cattle in the United States are regulated and approved by the Food and Drug Administration (FDA) as being safe for both humans and animals. Scientists and regulators are not perfect. They failed to see how the feeding of rendered carcasses to cattle might lead to mad cow disease. As this book is going to press a growth promoter used in the beef industry (Zilmax) is being accused of harming cattle health. At the present it is unclear whether Zilmax is the problem, but its maker is defending itself by remarking upon all the safety studies that were performed and reviewed by regulators. So, no, regulators are not perfect, but we know of no better way to determine what is safe and what isn't than the science regulators employ. We certainly trust the science more than rumor.

The European Union has a different opinion of growth hormones, and any beef the United States exports to the European Union must be hormone-free ("hormone-free" means the cow was not given synthetic beef hormones, as all food contains hormones). It is unclear why the European Union has taken this stance, given that the scientific literature and the World Trade Organization deem hormones to be safe. Some suggest it is just an excuse the EU uses to protect its beef producers against imports. However, the European Union does not allow its domestic producers to use synthetic hormones, so it is not surprising that this trade restriction was placed on the United States. European consumers seem to display more skepticism than Americans towards hormone use, and this ban probably just reflects different public attitudes.

If indeed children are maturing earlier, there are much better explanations for what may be contributing to it. Children who weigh more tend to mature earlier, for instance, and childhood obesity started rising around the same time growth hormones were adopted by the cattle industry. This is well known in the African country of Mauritania, where young girls are sent to a "fat farm" to gain weight (as being skinny is associated with poverty) so that they can better attract suitors, and so that they mature earlier.

If possible, at 8 or 9 years old she will begin to be force-fed until she prematurely matures into an adult woman.
—Aminetou Mint Ely, the Association of Woman Heads of Households in Mauritania, interviewed on "Winners & Losers," *Vice*, HBO. In the closed-captioning the word "prematurely" was placed in square brackets.

There is one legitimate reason to buy hormone-free beef, though. It has nothing to do with safety but everything to do with the eating experience. Cattle given synthetic growth hormones tend to produce tougher beef, so if you are willing to pay a higher price for tender beef, hormone-free beef may be the way to go. There is no such thing as a free lunch, though, as hormone-free beef will cost more and has a larger carbon footprint because of the longer time required to reach a slaughter weight.

Growth Hormones in Livestock Agriculture: Pork, Eggs, and Poultry

No growth hormones are given to hogs or chickens, mostly because they simply are not as effective as they are in cattle. So if you see pork, eggs, or chicken labeled as hormone-free, the seller is telling you the truth, but is trying to deceive you into thinking that its competitors do use hormones.

rBST Hormone in Milk Production

The hormone controversy is most intense in milk production. This is evident in the label on every bottle of milk in the United States. When a dairy cow gives birth, its pituitary gland begins producing the hormone somatotropin. This hormone diverts reserves of energy into producing milk. Dairy farmers can boost the cow's milk production by injecting it with additional somatropin. Manufacturing the hormone is difficult, though, and was not feasible until the Monsanto Corporation genetically modified a bacterium to produce rBST: recombinant bovine somatotropin. Now farmers can inject cows with rBST and produce more milk from each cow. This means more milk for each pound of corn fed, each gallon of water drunk, and each hour of human labor. Because resources are used more efficiently with rBST, it lowers the carbon footprint of milk.

Consuming milk from cattle that were administered rBST means humans are consuming a genetically modified growth hormone. Is that safe? The Food and Drug Administration (FDA) says it is, explaining that rBST is biologically indistinguishable from its non-GM counterpart (BST), that both are inactive in the human body, and any other differences between the two milks have no impact on human health.

Some have questioned the FDA's assessment, noting that since rBST boosts milk production by denying the cow of some of its reserve energy, the cattle might experience poorer health. Compare the body of a beef cow to a dairy cow and one will see the toll that high milk production can take (allowing some exaggeration, high-yielding dairy cows look like a hide draped over a bare skeleton with a huge udder underneath). Increasing milk production might compromise the cows' immune system, requiring the use of antibiotics. The resulting milk might then contain antibiotic residues and be less safe, the claim goes (though the FDA regulates antibiotic use to prevent this). If cows are negatively affected by the hormones, some wonder whether the consumers of the cows' milk will be harmed also. Although Monsanto claimed that rBSt had no impact on cattle health, when some of Monsanto's confidential data were made public by an anonymous FDA employee, the data suggested those claims were false. Cattle health was worsened by rBST. The arguments made against rBST then involve a conspiracy.

The conspiracy theory got even bigger. There was an official investigation. The FDA defended its position by publishing an article in the prestigious journal *Science*, but the conspiracy theorists (a term not meant disparagingly) observed that the main reviewer of the article had received compensation from Monsanto in the past. When an FDA employee was fired, some said it was because he expressed his belief that rBST needed more research before it should be deemed safe. The FDA was accused of manipulating data, or relying too much on Monsanto's honesty, and as a result failed to determine the extent to which rBST milk is safe (in regards to a substance called IGF-1).

So here is another case where a controversial issue requires one to decide whether to trust the regulatory agencies and the bulk of scientists, or whether to believe that the influence of a corporation is so pervasive that the truth is only being spoken by a few courageous journalists and scientists. As we have stated before, we tend to trust our fellow scientists and regulatory agencies, and so we are skeptical of the conspiracy theory. Some readers of the book *The World According to Monsanto* and viewers of the documentary *Ethos* might feel otherwise, and seek to avoid rBST milk. It certainly seems that the general public is more skeptical of rBST milk than our colleagues in agricultural colleges.

Have you noticed that milk often comes with the label, "This milk is from cows not treated with rBST," which is then followed by the statement, "The Food and Drug Administration has determined there is no significant difference between milk from rBST treated cows and non-rBST treated cows," a disclaimer recommended by the FDA to prevent the seller from being accused of false advertising? These seemingly conflicting statements reflect the desire of some farms to meet consumer demand for rBST-free milk and the FDA's belief that there is no legitimate reason for consumers to demand such a product. The FDA does not represent all of government, though. The Sixth Circuit Court of Appeals (ruling on whether Ohio should ban the labeling of rBST-free milk) concluded that rBST and non-rBST milk are materially different. The presence of the label saying the cows were not given rBST does stigmatize conventional milk, research has shown, making some people feel milk without the label unsafe. This is one battle food activists largely won, as most milk producers now prefer to sell rBST-free milk.

Consumer sentiment has opposed the rBST hormone with much greater intensity than synthetic growth hormones in beef cattle. Why? Probably because rBST involves genetically modified organisms (GMOs), something food activists particularly dislike. And the louder food activists shout, the more consumers listen to them. Although more than 90 percent of beef cattle receive synthetic hormones, that percentage for dairy cows is less than 25 percent.

Antibiotics and Livestock

You come down with a cold, so you go to the doctor. Most colds are caused by viruses, not bacteria, but some doctors prescribe antibiotics anyway. They won't make the cold go away. They won't make you feel better, save for any placebo effects. Antibiotics can only target bacteria, not viruses. So why do some doctors prescribe them? Probably because patients expect *something*, and will be unsatisfied if given nothing.

This is a common occurrence, and while doctors are simply satisfying their patients' wishes, they are also harming their health. This is because overprescribing antibiotics makes it more likely that harmful bacteria will develop resistance, making bacterial infections difficult to treat when they really do occur. In some countries one can even buy antibiotics without a prescription, making overuse of antibiotics even more widespread.

Antibiotics are also given to cattle, pigs, and chickens (not the ones laying eggs, though). Sometimes the animal is sick and receives a high dose. However, even if the animal is not sick it might receive low doses on a regular basis. These low doses are not enough to suppress an actual infection; nevertheless, they keep the animals healthier and growing faster. This practice may be good for pigs, but it poses health dangers to humans. Regularly giving livestock low doses of antibiotics is like presenting harmful bacteria with a weakened enemy. The bacteria "practice" fighting the weakened antibiotic, thus learning how to withstand a stronger dosage and eventually becoming immune. That resistant bacteria may later infect humans and thrive, regardless of the antibiotics the doctors prescribe. Or, even if the resistant bacteria do not infect humans they may share their resistance properties with bacteria that do (and yes, organisms can swap genes).

So antibiotics are overused in both humans and livestock. Doctors prescribe them at high doses when people are sick with viral infections, and livestock receive low doses regardless of whether they are sick. Both practices threaten human health, but it is unclear how serious the threat is, or which threat is greater. Doctors experience less criticism because they overprescribe antibiotics out of good intentions of improving human health, whereas livestock producers are seen to over-use antibiotics in search of higher profits.

Although it is clear that these low, regular doses of antibiotics given to animals can lead to human health problems, it is easy to overstate the problem. Around 80 percent of all the antibiotics sold in the United States are given to livestock, but most of these are not used by humans. The livestock category of antibiotics even includes ionophores, which have no human equivalent. In fact, ionophores are so different from other antibiotics that in 2007 Tyson Foods was able to raise chickens using only ionophores, and label their meat "raised without antibiotics" (something producers are no longer allowed to do). Although 80 percent of antibiotics sold are given to livestock, if we look only at the antibiotics used by both animals *and* humans, that percentages falls to 45 percent.

To what extent is human health threatened by antibiotic use in livestock? People disagree. One view suggests the probability is low. For regular, low doses of antibiotics to harm human health the following chain of events must take place: (1) a bacteria infecting livestock must develop resistance to an antibiotic, (2) that antibiotic must also be used for humans, (3) the resistant bacteria must also be able to infect humans, and (4) they must make people so sick that they need antibiotics. To some observers, the probability of all four events occurring seems incredibly low. A researcher in the animal health industry (an admittedly biased source) calculates this probability to be as low as 0.00034 percent, and some scientific articles suggest a similarly low probability.

This probability makes a number of assumptions that might not be true, and when those assumptions are relaxed, the probability of a human health threat rises. One assumption regards the horizontal transfer of genes. However odd it may sound, bacteria can share genes for antibiotic resistance. This means that if one bacterium develops resistance, even if that bacterium cannot harm humans, it can share its immunity with bacteria that can. Though the rate of horizontal transfer is unknown, it has been established that antibiotics in hog feed increase gene transfer.

How do infections spread from animals to humans? It isn't just through the animal's meat. Using manure as a fertilizer can contaminate vegetables. Farmworkers can carry the bacteria with them as they leave the farm. Food activists love to tell the story of Russ Kremer, a swine farmer who routinely gave his hogs antibiotics at a low dose, only to personally acquire an antibiotic-resistant infection after his skin was pierced by a boar. Kremer is now a spokesperson for the movement to curb antibiotic use in pigs. Is his experience highly unlikely, or does it portend trouble? Hard to tell, but studies have found that workers on farms using antibiotics do carry antibiotic-resistant bacteria with them as they leave the farm, while those working on antibiotic-free farms are less likely to do so.

There is a benefit to antibiotic use in livestock, other than lower meat prices. Humans can become sick from all bacteria, not just those resistant to antibiotics. If hogs are healthier when they receive regular, low doses of antibiotics, then there will be fewer bacteria overall, and perhaps fewer human infections. For instance, one study found that hogs raised in an antibiotic-free setting were infected with salmonella at a greater rate than hogs on farms regularly prescribing antibiotics.

Conceptually, it is impossible to say whether antibiotic use in cattle, pig, and broiler production heightens or reduces human health threats. It is an empirical question, which requires one to guess the likelihood that bacteria will transfer genes, horizontally, between each other. Many scientists and most health organizations conclude a real human health threat exists and that antibiotics should only be used therapeutically and at appropriately high dosages to treat observed infections, and many held this opinion even in the 1970s. The American Medical Association openly opposes the regular use of antibiotics, and the World Health Organization considers antibiotic resistance one of the top three threats to human health. The European Union went so far in 2006 as to ban the use of antibiotics in livestock, unless the animal is sick. The FDA in the United States has thus far been reluctant to follow the European Union, but in 2012 it initiated new rules that seemed to be heading in that direction, and at the end of 2013 began taking action to make sure antibiotics used to treat humans are not also used in livestock production. Because the livestock industry has already transitioned away from using antibiotics that are also used by humans, these changes by the FDA are expected to impact the livestock industry only slightly.

The livestock industry continues to argue that the benefits of the antibiotics are worth the cost, and that claims the antibiotics are creating "superbugs" (human bacterial infections that are resistant to all antibiotics) are not founded in empirical evidence. Though it might seem the industry is acting like a "merchant of doubt," even the scientists blaming superbugs on livestock admit that the impact on human health may be impossible to measure scientifically.

Are we "eating" antibiotics and antibiotic-resistant bacteria, when consuming a product not labeled "antibiotic-free?" Well, it is unlikely to encounter an "antibiotic-free" label, especially for regulated products, because all food produced according to law is virtually free of antibiotic residues. The USDA prohibits antibiotics from being used close to slaughter, and the definition of "close" is set to ensure all antibiotics are given time to clear the animal's body. Because some people are allergic to antibiotics, residues on food must be close to zero to prevent allergic reactions. The USDA tests products for antibiotic residues, and only rarely do they exceed their maximum threshold set by the government. The vast majority of the time antibiotic residues in meat, eggs, and dairy products are exactly zero. Some products may say "Raised without antibiotics," which means the animal was never given antibiotics. While this may initially sound appealing to some, giving farmers a premium for not using antibiotics means they are more likely to let a sick animal go untreated—that's bad for animal welfare. The restaurant chain Chipotle in 2013 sought to strike a reasonable compromise, where it would accept meat from animals given antibiotics, so long as the antibiotics were given in response to a clear sickness. There is little reason to be concerned about antibiotic residues in food. Antibiotic-resistant bacteria in food is a greater concern. In April 2012 an alarming report found that most turkey, pork, and beef (and close to half of chickens) sampled from supermarkets did indeed contain antibiotic-resistant bacteria. A few months earlier *Consumer Reports* announced it had sampled pork products and found many contained bacteria, some of which were resistant to antibiotics.

Though they sound scary, the stories are a bit deceiving. Here are several things to consider. First, you can find antibiotic-resistant bacteria anywhere, including furniture, your navel, and your nose. Second, most of the bacteria found on pork were a strain called *Yersinia enterocolitica* that the USDA does not test for, because even the most scientific tests results in lots of false positives (i.e., the tests say the bacteria are present when in reality they are not). Third, many of the antibiotics the bacteria were resistant to are not used by humans.

If those two reports still seem scary, one can purchase food from animals raised without antibiotics (including organic food), where the animals probably harbor fewer antibiotic-resistant bacteria. They might, however, contain more bacteria overall.

Though it remains a controversy, both the United States and the European Union are taking measures to reduce antibiotic use in agriculture. What would happen if it were banned in the United States? To gauge the impact of such a move we can look to Denmark, where in 1995 a ban was placed on routinely giving antibiotics (except if the animal was sick). At first, total antibiotic use rose, as animals became sicker and required more antibiotics to cure infections. However, the overall dose per pig has since fallen and there are fewer antibiotic-resistant bacteria in Danish meat than in the meat Denmark imports. Some say the Danes have experienced health benefits from the ban, while others claim that any benefits have been offset by a reduction in protein consumption caused by higher pork prices.

Denmark has also taken the lead in collecting detailed data on how antibiotics are actually used on the farm, allowing their scientists to identify the types of antibiotics and settings that encourage resistance. The United States only publishes numbers on the total amount of an antibiotic used. This makes it difficult to craft a coordinated response to the threat of antibiotic resistance.

According to most writings and documentaries by food activists, the use of confinement facilities for livestock is only possible if the animals are constantly given antibiotics. They suggest that without antibiotics the livestock industry would evolve into small farms using more "natural" production methods. Though the claim seems logical, it is an empirical question, and this did not happen in Denmark. In fact, during the period in which the ban took effect, bigger farms replaced smaller farms, and the pork industry retained the use of confinement facilities with only minor adjustments.

9

PARTING THOUGHTS

News at *The Onion* may be fake, but it often speaks a deep truth about society. In September 2013 *The Onion* ran an article with the headline "Guy Looking to Feel Horrible about Aspect of Everyday Life Decides to Watch Documentary." The character interviewed in the article remarks, "I already feel terrible about American politics, advertising, water, dolphins, fast food, and Walt Disney, so let's see what other documentaries can make me feel terrible about something it never occurred to me to feel terrible about before."

Notice the character included fast food. Over eighteen documentaries are available on Amazon and Netflix telling the audience how industrial agriculture is poisoning the soil, torturing animals, and sickening the public. This at the same time modern democracies have access to cheap and nutritious food like never before (whether we *choose* healthy food is another matter). To some, the documentaries (and books of a similar spirit) point to real problems in agriculture, but to others they are the mere manifestation of "muckrakers" seeking fame and money.

It was largely these documentaries that provided the motivation to write this book. We noticed that the public is more interested than ever about how its food is raised, but scientists are reluctant to engage that interest. Their reluctance is understandable. Regardless of what they say, somebody is probably going to be upset, and scientists enjoy learning more than they do debating. However, watching the food documentaries, we realized that, regardless of whether one agreed with the content, they were asking good questions—many questions the scientific community had ignored.

As we researched the controversial issues in agriculture we gained an appreciation for controversy itself. If there is one thing readers should take from this book it is the importance of good government in regulating things like pesticides, GMOs, synthetic growth hormones, and the like. Conventional food is considered safe and healthy by the authors because we place considerable confidence in agencies like the Environmental Protection Agency, the Food and Drug Administration, the Department of Agriculture, and their European counterparts. Watch the aforementioned food documentaries and it will become apparent that most food activists feel differently. Yet it is this skepticism about regulatory agencies that helps the agencies perform so well.

Food activists may be constantly looking for any reason to criticize agriculture, and sometimes the criticism is unfair, but if nobody is looking for problems like water pollution they will not be recognized until the consequences are terrible. Consider China, where the absence of social activism allowed firms to irrigate rice with polluted waters, going unnoticed until 10 percent of its rice was contaminated with cadmium. Problems are best solved if recognized early, and even if food activists seem a little too eager to be in the vanguard, their enthusiasm serves a useful role. It took the activism of Rachel Carson to make us aware of pesticides' potential dangers, and consequently, conventional food is safer. It took the activism of animal welfare groups to make us think harder about the well-being of chickens. As a result, groups like the United Egg Producers have voluntarily improved their cage facilities. Industry groups were not the originators of the sustainability movement—activists were—but industry groups now measure their carbon footprint and seek ways to make it smaller.

Controversy is the pulse of a democratic society, but it is not a peaceful pulse. The assembly in ancient Athens may have been democratic but it was not pleasant—debates were tense and fiery. Likewise, modern democracies wage turbulent public relations battles over agricultural controversies, exchange petty insults on websites, and hire lobbyists as mercenaries in the fight over farm bills and environmental regulations. Words exchanged are sometimes ridiculous, like when Maria Rodale wrote to President Obama saying we are "no better" than Syria because we also use chemical weapons on our own people in the form of pesticides. Or when the livestock industry suggests that the well-being of farm animals can be measured solely by its profitability—that a profitable pig must therefore be a happy pig. Yet one can't have a real debate without extreme comments being made, and such comments should not dissuade others from contributing their more moderate views.

Even though no side ever "wins" these battles, the debate makes us constantly reevaluate how we produce food with the objective of making it greener, safer, healthier, and more abundant. Debates may at times be destructive and lead us into erroneous beliefs, but societies without social controversies are not utopias, but dystopias (think North Korea). In the course of writing this book we found ourselves asking new questions, and learning a good deal, the more seriously we took food documentaries. We discovered that we hadn't really thought much about the long-term consequence of relying on chemical fertilizers, and that we knew very little about how pesticides and genetically modified organisms were regulated. Moreover, we learned that we had far fewer "answers" to the important questions than we initially believed. Engaging these agricultural and food controversies taught us much, and made us better researchers and teachers. So let the food debates continue, regardless of where they lead. They are better than no debate at all.

INDEX